The Future of

About Hyperloop Advanced Research Partnership

Hyperloop Advanced Research Partnership (HARP) is a 501(c)(3) non-profit organization incorporated in 2017. The HARP Mission is

To promote the collaboration, research, funding, and knowledge sharing necessary for the development of high-speed tube transportation networks and standards around the world.

HARP relies on our enthusiastic members and volunteers. Join us at HyperloopPartnership.org

HYPERLOOP
The Future of High-Speed Transportation

Hyperloop Advanced Research Partnership
HyperloopPartnership.org

Editors and Contributing Authors

Stephen A. Cohn, Ph.D.
National Center for Atmospheric Research, retired

Mary Ann Ottinger, Ph.D.
University of Houston

Thierry Boitier, M.Sc.
TransPod Inc.

R. Richard Geddes, Ph.D.
Cornell University

Ian Sutton
Risk Management Consultant and Author, Sutton Technical Books

Brad Swartzwelter
Amtrak

Dario Bueno-Baques, Ph.D.
University of Colorado Colorado Springs

Radu C. Cascaval, Ph.D.
University of Colorado Colorado Springs

Dane S. Egli, Ph.D.
Department of Homeland Security

Denver, Colorado

HYPERLOOP, The Future of High-Speed Transportation

ISBN # 978-0-9725955-0-6

Published by **Alder Press 2021**
Denver, Colorado, USA

Cover and models designed by:
Woolf&Woode Design
consultancy
www.woolfandwoode.com

Dedication

**To the visionary transportation leaders of history,
of the present, and of the future.**

Contents

Preface

In 2016, members of the newly-formed organization, Hyperloop Advanced Research Partnership (HARP)—an independent group seeking to inspire next generation high-speed tube transportation—made visits to senior stakeholders in Washington, DC. The purpose of this outreach was to inform key decision-makers, gauge the level of interest, and recruit potential support for what was an uncertain leap of faith. The White House, Pentagon, Department of Transportation, Capitol Hill, and Transportation Research Board (TRB) all expressed the same circumspect interest when we presented the potential utility—strategically and operationally—of future high-speed transportation systems, captured in one word, "hyperloop."

Their questions ranged from, "What is it, how does it work?"; "How much will it cost?"; "How soon will it be operational?"; "How will it integrate with existing transportation infrastructure?"; And there were skeptical comments such as, "The science and technology may work in theory, but there has been no successful prototype testing;" or "Even if there is a proof-of-concept, existing intermodal transportation organizations will be threatened;" "While the science, engineering, and technology may be in place, you will never overcome the political, policy, legal, and right-of-way hurdles." Imbued with a sense of hope, our team finished the week with a meeting on Capitol Hill, which included some young staff members from the House Committee on Transportation. One particular staffer, listened carefully, and provided some timely words of encouragement, "We know there is a better way, but we just don't know what it is…" This book is about the "better way."

Many nations today are beset by congested transportation networks, unstable and highly-interdependent supply chains, antiquated intermodal systems, emerging technology questions, environmental challenges, asymmetric security threats, and a global pandemic that has introduced disruptive uncertainties into all markets. Further, airport runways, transcontinental highways, regional airports, train railways, maritime ports and waterways— the vital ligaments of economies—are unable to reliably and rapidly move people, cargo, and commodities. These transportation systems are existential necessities upon which any nation depends upon for safety, security, and economic survival. Indeed, they are the sine qua non—essential ingredient—of organized society. This book highlights distinctive features, designs and capabilities, of transformational tube transportation systems that will add to existing modes in a complementary and integrated way—with greater speed, capabilities, technology, capacity, and resilience.

Over the past five years, HARP has been a harbinger for high-speed tube transportation by convening conferences—in Denver, Houston, New York City, Washington DC, Los Angeles, and at the Colorado School of Mines in Golden, CO— attended by transportation futurists, students, policymakers, and researchers from the public, private, and academic sectors. The commercial industry, start-up companies, student pod-competition teams, and transportation innovators are clearly making the investments necessary to move this idea from concept to reality. This book provides a catalyst by defining standard terms, identifying technology gaps, and asserting vital design breakthroughs; and will inspire tangible progress in hyperloop transportation research, development, and education—a faster, safer, cleaner and better way.

Work on this book started early in the year 2019. From the beginning it has been a team project. Each chapter has a primary author, but all authors contributed to the overall effort. Writing a book in this way is indeed like herding cats, but it does mean that the book is a multi-disciplinary effort. Our collective expertise includes biology, industrial health, passenger train operations, physics, environmental science and industrial safety. We have looked at hyperloop from many points of view.

HARP is a non-profit organization made up entirely of volunteers who have given freely of their time and expertise to help us all understand hyperloop: what it is, and what it might mean to transportation of the future. None of the volunteers have received any remuneration — indeed many have given not only their time, they have provided funding for the project. Similarly, revenues from the book will go entirely to the HARP organization.

Wherever possible, we have worked with those companies that are building commercial hyperloop systems. Their input has been invaluable. However, HARP does not endorse any particular company or technology — our interest is entirely generic. We also see it as our responsibility to identify problems and challenges. We support the hyperloop concept, of course, but we are not cheerleaders.

Aspects of hyperloop involve the use of new technologies, many of which are still being tested. This means that the authors of this book do not have the final answers — like everyone else, we are on a learning curve. Down the road it is likely that we and others will need to pivot as the industry learns and moves forward. At the same time, we wanted to get the book and current thinking into circulation. Three quotations come to mind. The first is, "Shoot the engineers", the second is, "Perfect is the enemy of good enough", the third is "Enough is enough already". There's a good time to stop writing — and that time is now. If we waited for answers to all questions this book would never have been published.

We encourage you to share your opinions with us. Please tell us what you think about this book, and about hyperloop in general at https://www.hyperlooppartnership.org/contact.

CHAPTER 1. Hyperloop: Spirit of Exploration
Primary Author: Dane S. Egli

Since Elon Musk published the Alpha white paper in 2013—asserting the feasibility and advantages of hypersonic high-speed tube transportation—there has been a flurry of interest, reflected in published research papers, operational tests, feasibility studies, technical demonstrations, student pod competitions, and exciting investments among public, private, and academic "explorers." These transportation pioneers—like their technological predecessors—have been captured by the uncommon spirit and success of Henry Ford, the Wright Brothers, Robert Goddard, and Charles Lindbergh, all of whom demonstrated the patience and endurance needed to navigate courageously toward a new frontier. This book is dedicated to the explorers—especially the students, transportation futurists, and commercial companies—who are cutting new trails with courage and vision.

In this case—similar to our historic forerunners—the landscape ahead holds unique uncertainties and the challenges of introducing a new infrastructure system into an already-complex network of rail, roads, rivers, runways, ports, and pipelines. Truckers, commuters, commercial maritime shippers, the airline industry, and market investors are paying a penalty for congested tracks, highways, airports, waterways, shipment delays, and disruptive natural and man-made disasters. The vision of hyperloop addresses these, yet there remains uncertainty about the reliability and marketability of this nascent and as-yet-unproven transportation technology.

The major transportation breakthroughs of the past two centuries brought historic efficiencies and economic growth. They enabled consequential modernization in military sciences, workforce opportunities, industrial technologies, and market-trade conveyances. When introduced, disruptive new forms of transportation challenged the status-quo and faced delays and resistance from traditional legacy markets. The first railroad was granted a charter long before railroads were commercialized; the transition from merchant sailing ships to steam-powered vessels was uncertain, forcing steamers to maintain canvas sail lockers even after propellers were installed; automobiles challenged horse-and-buggy companies; the airplane success on Kitty Hawk, North Carolina sand dunes was not initially embraced in America, causing the Wright Brothers to launch their enterprise with European investors. The years since these transportation bellwethers emerged have been marked by improvements in their design and efficiency, as well as progress into space travel. The time has come to add another transportation capability to the intermodal system of systems. Like the transportation inventions of the past—that advanced the national and economic security of nations—this transportation modernization will infuse strategic benefits to societies and nations.

The proposition of this book is that high-speed tube transportation represents a quantum leap forward by transforming speed, range, capabilities, and resilience of terrestrial transportation. Fundamental imperatives remain to be addressed: (1) Is there a commercial market and government sponsors prepared to make the necessary investment to advance this innovation?; (2) Will the science, technology, physics, and engineering of this design be tested and evaluated to an extent where industry and government can confidently support it?; and (3) If commercialized, can the sponsors and owners gain the political, policy, and legal clearances for land rights-of-way, technical standardization, and environmental requirements? These are essential ingredients to a successful market analysis and investment proposal as the enterprise seeks validation and consensus to chart a realistic path forward.

Forging a resilient consensus within a company, city government, or commercial enterprise is a daunting task. The introduction of a new transportation system will require forming that consensus within an international community of nations facing unstable economic markets, uncertain commitments to rules-based order, persistent asymmetric threats, and a global pandemic that poses an existential threat to some societies, markets, and businesses. Against that backdrop of geopolitical uncertainty, emerging technologies, and globalization, the promises and potential of hyperloop – that it will be safe, clean, fast, affordable, and provide a potential remedy to these sobering challenges of the

21st Century – should make it a priority for people and nations. In times like this there is a clear need for national—and international—strengthening of security, safety, economic, and environmental resilience by improving and modernizing technological capabilities across all domains—air, land, sea, space, and cyber. Hyperloop can transform our transportation in ways that will support a strong and interconnected future.

In this book, authors from many disciplines have combined efforts to describe and discuss aspects of hyperloop and reimagine the future of transportation. Not that it will replace the existing complex interdependent critical transportation infrastructure systems—Chicago trucking firms, Houston refinery pipelines, Los Angeles maritime port shipping, Atlanta airport services, Dubai intermodal transit network, Rotterdam container hubs, or Singapore high-speed rail—but that hyperloop can be introduced in a way that will enhance and reinforce existing transportation systems by integrating complementary capabilities and capacities. Navigating the landscape ahead will require several indispensable qualities—common threads running through every chapter of this book—for hyperloop systems to move beyond the aspirations of a few well-financed companies; the excitement of university student clubs and pod competitions; and a non-profit industry support organization like HARP. It will require *unity of effort*, *visionary leadership*, and a *clear value proposition*.

Unity of Effort

Unity of effort is the collective action among a dedicated group that recognizes that some things—clean air, clean water, economic and military security—are public goods and important enough that their common benefits outweigh the needs of any single consumer or market participant in "governing the commons." An underlying premise of advancing hyperloop technology must be that transportation is a shared resource, where societies contribute for the good of all to ensure a new and better outcome. If the companies and municipalities building tube transportation are willing to participate in a collaborative "shared economy," there will be the prospect of an operational system from New York City to Washington DC, or Dubai to

Abu Dhabi, or from Los Angeles to San Francisco that reflects the collective design of many contributors as a valued enterprise. For example, consider the inventions and number of separate patents that are part of an automobile—everything from the transmission, electrical wiring, suspension system, fuel supply design, and even the exhaust system are derived from the collective inventions, engineering, and best practices of individual sources that contributed synergistically to the complete product. Development and realization of a complex hyperloop network will require a similar unified and adaptive approach, with collective action and unity of effort.

Visionary Leadership

Visionary leadership is the indispensable ingredient of any major breakthrough in a whole-of-society effort that requires the foresight and proactive efforts of empowered leaders. The integration of a new transportation system depends on integration not only of infrastructure but of stakeholders. Garnering the acceptance and support of the many unions, congressional delegations, and financial players that are heavily invested in the current system—and in its sustained unchanging operation—not only requires a unity of effort across private industry, but also enlightened and innovative leadership. The success of hyperloop development and achieving its benefits will require this leadership. One strong model can be the dynamic collaboration of Public, Private, Academic Partnerships (PPAPs), where good governance, financial profits, and thought-leadership meet in a market environment-demonstrating the operational utility of a new and profitable form of transportation.

Value Proposition

Value proposition is the central factor in determining hyperloop transportation development and the necessary transfer of technology to an integrated infrastructure project. If independent review of this emerging technology substantially validates it promises, safety, cost, and reliability, there is still the need to attract market investors - it must show the potential for a commercial return-on-investment (ROI). While feasibility studies; 500-meter field prototypes; technology demonstrations; operational proof-of-concepts; research & development (R&D) evaluation reports; operational tests and evaluations; and exercises are important steps to establish its technology readiness and operational utility, there is no substitute for a vetted projection of financial profits to attract the attention of private-equity investors, venture capitalists, start-up entrepreneurs, government agencies, and congressional interests. During the early stages of a breakthrough innovation in any industry, the most common questions asked by members of the public and private sectors are in this area of economic viability, potential commercialization, and monetization of cost-benefit estimates.

The excitement of speed, technology, clean transportation, and innovation has unleashed a spirit of exploration, and has attracted a global community of transportation developers, planners and enthusiasts. This spirit was heard it in the presentations of young engineering students at HARP's LoopTransPort2018 conference at the University of California-Los Angeles (UCLA), on social media videos made by leaders of private hyperloop companies building a test track in France, during the SpaceX pod competitions in California, and when a State Secretary of Transportation confidently announced his governor's approval of a provisional permit to develop a track from Baltimore to Washington DC. For an endeavor like this to get started, eagerness and brave optimism must be joined with visionary leadership, effective unity of effort, and a measurable value proposition. The following chapters attempt to frame those ambitious goals by introducing hyperloop technology as a catalyst for the future of transportation.

CHAPTER 2. Why Hyperloop Matters
Primary Authors: Ian Sutton and Stephen A. Cohn

Visions of a World with Hyperloop

Just as the advent of automobiles and commercial aircraft transformed societies the world over, travel at the speed promised by hyperloop can reshape our communities and daily life. Imagine the opportunities available in a world with hyperloop transportation. Each of the following visions of hyperloop companies or enthusiasts point to its far-reaching and diverse potential.

Live in Paris, work in Cologne. Because of hyperloop's speed, people have much greater freedom in choosing where to live relative to their daily workplace. A 30-minute hyperloop ride into a city center could begin several hundred miles away rather than a few tens of miles. Someone living in Paris, for example, could easily commute to Amsterdam, Cologne, or Bern daily. Similarly, while living in Hyderabad, one could reasonably commute to Bengaluru or Pune, or maybe even Chennai. Such facility of movement is bound to change the shape of cities and suburbs in unforeseen ways.

Riding in comfort. The experience and amenities of riding in a hyperloop pod remain to be decided, but some features are common to all proposed visions. Magnetic levitation and propulsion will give a very smooth ride during the glide phase," that is, between initial acceleration and final deceleration. Dashboards and simulated windows will give passengers a sense of time and the trip's progress. The experience at stations, and upon entering and exiting pods, should also be seamless as this technology helps to remove the choke points of today's airports.

Faster door to door service. As transportation becomes more and more autonomous, hyperloop can be a lynch pin of a seamlessly integrated, multi-modal transportation network that gets people to their final destination without delays and scheduling complexity. Hyperloop covers long distances rapidly, while other technologies will address the "last mile" of getting people and packages efficiently from door to door.

Promoting a stable climate and clean air. Hyperloop will have less impact on climate change than earlier methods of mass transportation. Today, about one quarter of all the CO_2 released into the atmosphere by humanity comes from modern transportation. By substituting hyperloop for some of the trips that today are made with gas, diesel, and jet fuel, hyperloop becomes an integral part of the global transformation away from the harmful burning of fossil fuels. Moreover, it will not contribute to the smog and particulate pollution so much of the world's population now breathes each day.

Faster package delivery and improved supply chains. Transport of cargo may be the economic driver of hyperloop development. In the 1990's companies such as FedEx, DHL, and UPS found there was a strong market for rapid, reliable package delivery. Also, manufacturing companies built just-in-time delivery into their supply chains to save the expense of local warehouse space and logistics. Hyperloop adds value to both these innovations in delivery. It will increase the reliability and flexibility of package delivery to peoples' homes, as well as improve the reliability and speed of manufacturing supply chains.

Invisible infrastructure. Building hyperloop in underground tunnels makes sense for many reasons. It insulates the infrastructure from large temperature swings that cause materials to expand and contract, addresses the system's physical security, enhances the longevity of the infrastructure, and reduces concerns over disruptions to neighborhoods. An underground tunnel configuration also solves perhaps the largest challenge for hyperloop – how to create the long, straight rights-of-way needed for high-speed travel. The flip side to this approach is that the time and cost of constructing such tunnels is much greater than for constructing surface infrastructure. Time will tell how much of hyperloop is invisible.

More beaches to play on. Today's large cargo ports take up miles of valuable seashore as huge container ships are loaded and unloaded, and cargo is sorted and moved to and from all other modes of transportation. With hyperloop, ports in the near future may be very different. Imagine ports where cargo is unloaded at floating offshore docks, then moved with hyperloop to sorting and transfer facilities well inland. This bypasses the shore itself, freeing miles of beach and shallow harbor for recreation, retail, or residential development.

Imagine replacing port facilities with beach recreation (Madagascar, Singapore).

All of the afore-mentioned visions are possible, if societies choose them. Many challenges remain – in engineering, in finance and funding, in regulation and safety, and especially in public acceptance and political will. Questions also arise about the true final shape of this new transportation, for example, the real speeds, energy efficiency, and convenience achieved. Yet one critical aspect holds firm –the laws of physics do not limit hyperloop's creation. It may be seen as next in a series of milestones that mark the evolution of transport in the United States.

Hyperloop as the Next Evolution in Transportation

Every so often technology and human ingenuity spur a fundamental shift in transportation for people and freight. Consider the following historical examples from the United States, each of which transformed life for the entire nation.

In a key moment of national transportation history, tracks from the Central Pacific and Union Pacific railroad companies were connected at Promontory Summit, Utah in 1869. From then on people could travel coast-to-coast in just a few days rather than weeks. Railroads greatly expanded trade and spurred economic development.

In 1956 President Eisenhower authorized the Federal-Aid Highway Act. This governmental initiative led to the creation of the interstate highway system, now 46,000 miles (70,000 km) long. The system contributed to why there are now more cars in America than there are people over 18 years of age. The age of highways led to the growth of suburbs, satellite cities, and further expansion in the reach of commerce.

Just a couple of years later, in 1958, the introduction of the Boeing 707 ushered in the jet age. Air travel was no longer a prerogative of the rich. Families and businesspeople could now routinely travel to distant locations at a reasonable cost.

Transportation is entering another time of transition. Established modes have run into the all-too-familiar limits of congestion, longer journey times, and insufficient space for expansion. Their environmental impacts have also become onerous. A fundamental change is needed. The new form of transportation known as hyperloop has the potential to be that change. It can be fast, convenient, and energy efficient. Powered by clean electricity rather than fossil fuels, it will also be more environmentally friendly than earlier methods of transportation. The emission-free operation of hyperloop will be a critical characteristic of the 21st century.

When Will Hyperloop Be Here?

While excited speculation and company claims abound regarding when the first commercial hyperloop line will carry passengers, the reality is that this enterprise will take time. The effort does have many things in its favor. Companies working toward hyperloop have brilliant engineers and are backed by enough investment capital to make progress. Governments around the world are fully engaged in feasibility studies. Yet the scope of creating national and international networks of hyperloop lines compares to that of developing highway systems, rail networks, and jet travel. Just looking at the timelines for such networks illustrates the many years it takes for these systems to be fully developed and widely implemented.

Although railway lines were being laid in the United States in the 1830s, it took 40 years to complete the transcontinental railroad, and it took many more years to establish a true national network. Automobiles were being built in the 1890s, but mass production only started with the Ford Model T, introduced in 1908. It would take another 60 years before freeways connected cities across America. Although a primitive jet engine was tested and flown in 1940, it wasn't until the 1970s that the jet age fully arrived.

These historical examples provide some guidance for the time it might take for a hyperloop system to be implemented at national and continental scales. First, it seems to take a decade or so for new technology to be intensely developed and accepted, followed by infrastructure and networks developed over several decades.

The hyperloop industry is not yet at the point where the technology is established. Nor have basic standards, such as the diameter of the tube, been established. The first commercial system may be a few years away, but a fully developed hyperloop network will take decades. Beyond the technological challenge, the speed of its development will depend on public demand, economics, and political will.

Things to Consider

Freight and/or passenger transport. Most of the discussion to do with hyperloop assumes that it will be used for passenger transportation. Yet its principal benefit may turn out to be the fast transportation of freight and goods, particularly containers and packages. Throughout the world, railway systems for transporting passengers require government subsidies. Hyperloop will likely be no different. However, freight transportation's typical profitability and financial self-sustenance make subsidies unnecessary. If hyperloop can carry freight profitably, then many of the financial concerns associated with this new method of transportation would be eliminated.

Freight's inherent safety also offers advantages over passenger service applications of hyperloop (see Chapter 7). With no people on board, no one would be injured or killed in the event of a high-speed crash. Beyond considerations of profits and safety, the choice between freight vs. passenger transport includes fundamental choices to be made early in the design process. For example, freight packaging configurations could impact the size of tubes or tunnels.

Cost and revenue considerations. At this stage in its development, no reliable estimates for the cost of creating or maintaining hyperloop systems are available. In fact, some of the basic assumptions that affect the cost have yet to be made. For example, infrastructure placed on pylons, or at grade, or below ground in tunnels will all have very different construction costs and maintenance needs. The decision to build a large tube capable of transporting full-size hi-cube freight containers or containers used for aviation freight will be more expensive than to build a smaller one adequate for passengers. And amenities for passengers might also affect the operational cost.

Hyperloop economic considerations include not only operation and maintenance costs, but also the revenue side. Revenue models for passenger transportation vary greatly. Trains and bus networks, which are considered a public good, have been highly subsidized, while private, independently-profitable airline, ride-share, and taxi services exist alongside. Similarly, postal services are often subsidized, while private freight transport has been reliably profitable. The advantage offered by an ability to move bulk freight at airplane speeds increases the chance that the hyperloop enterprise would be profitable. Thus, hyperloop companies, so far, are proposing that most or all of the financial risk for hyperloop be taken by the private sector (i.e. venture capital investors or the financial markets).

Environmental considerations. Hyperloop's ability to run without burning fossil fuels, an advantage over some existing transportation, has also led to support. The support relies critically on the commitment of operators to source the electricity that powers hyperloop from renewable rather than fossil fuel sources. Beyond energy consumption during operation, the construction of transportation itself can use large quantities of fossil fuel. Consequently, emissions from hyperloop, as with any infrastructure project, should be considered on a full life-cycle basis.

Safety and security considerations. People and societies value safety, and they rightly insist that products and services be safe, with a low risk of injury. Yet we hear regularly in the news about automobile accidents, train derailments, and even aircraft crashes. Actually, aviation has an excellent safety record, and its attention to safety is reinforced by regulations, guidelines, and a purposefully-developed safety culture. The hyperloop industry can learn from this and commit to building the safest transportation network possible, both from accidental and intentional sources.

Travel in an age of pandemics. In late 2019 and 2020, as this book was being written, the pandemic of COVID-19, the disease caused by the new SARS-CoV-2 coronavirus, led public transportation to be severely scaled back, and even automobile use diminished. The design of any new system should leverage an ability to learn from the past and integrate new ideas. The full range of precautions that should become routine to address contagious disease remains undeveloped – although measures like automatic screening for symptoms, changes to the way air is circulated and filtered, and improved lavatory safety have all been suggested. As more is learned, the design phase of hyperloop should consider how this aspect of safety can be integrated from the start.

A TGV Duplex high speed train in 2018. In 2020 France converted TGV trains into ambulances to transport critically ill coronavirus patients across the country.

CHAPTER 3. How Hyperloop Works

Primary Authors: Dario Bueno-Baques and Radu C. Cascaval

Introduction

Hyperloop transportation relies on a simple concept: build a system in which vehicles can travel at very high speeds in a reduced pressure environment. This approach was proposed as early as 1799 by George Medhurst in London. Called a Vac-Train, it never went beyond a patent, nice drawings, and some public excitement. In the two centuries since Medhurst, several similar projects have been proposed, each introducing new technologies to realize the concept. Over time great engineering progress has since been made. For example, magnetic levitation and non-contact electromagnetic propulsion have been introduced. Some projects, like the Swissmetro of the early 2000s, received greater attention and were successful to the "proof of concept" phase. But sadly, beyond the technical challenges, economic and political hurdles have stalled such projects.

The 2013 Alpha paper published by Elon Musk and SpaceX became the foundation to re-ignite the ideas behind fast tube-based transportation. Its vision of incorporating high speed, great energy efficiency, low cost, and convenience turns out to be a tough design challenge. Although the technical challenges are still substantial, rapid advances in technology suggest that solutions likely will be found. How long it will take to develop, test, and perfect these solutions remain unknown.

This chapter provides a glimpse into the physics and engineering behind the hyperloop concept. It returns to first principles and explains what technologies may hold the key to realizing the vision of high-speed tube transportation in the near future. Concepts like attaining high speeds at high efficiencies and low energy usage are interdependent and thus considered in detail.

The Future of High-Speed Transportation

Working Within the Laws of Physics

Great technical challenges to the development of hyperloop stem from its goal of achieving high speed, while maintaining the energy consumption at reasonable levels.

The first challenge is to reduce the air resistance, also known as aerodynamic drag. Pushing anything rapidly through the atmosphere requires a large amount of energy. Hyperloop addresses this issue by introducing a reduced-pressure environment. The use of a tube to control the environment outside the vehicle stands out as a unique feature of this new transportation mode. Air pressure can be lowered by evacuating air within the enclosed tube. While this reduces resistance, the constrained space creates another challenge: even at low pressures, the air confined in the tube must flow around the hyperloop vehicle. This confinement creates aerodynamic drag, which increases the energy required to maintain forward motion. The higher the speed of the vehicle, the more this effect gets amplified, reaching a limiting speed where the flow becomes 'choked', demanding a large energy expenditure. This important effect is known as the Kantrowitz limit and is discussed further below.

A second challenge to achieve high speed is to reduce, or eliminate completely, the contact between the vehicle and the track. Car tires and train wheels use friction to propel a car forward. At very high speeds however, increased friction can make a vehicle become unstable. To address this problem, the Space-X Alpha paper proposed "air bearings" (basically a cushion of air) to lift the vehicle off the track. The hyperloop vehicle transport would then be analogous in motion to an air hockey puck. Alternatively, to reduce friction most current designs lift the vehicle off the track by means of magnetic levitation. Although magnetic forces can be used to achieve contactless motion, this itself presents considerable technical challenges, such as the magnetic drag. Furthermore, a contactless motion requires ways to propel and brake the vehicle. Current hyperloop designs favor the use of an electromagnetic propulsion as it is considered the most feasible solution for this challenge.

A third challenge is because the hyperloop must not exceed an acceleration that maintains a physical comfort level for the passengers. Most people find a reasonable comfortable maximum acceleration to be around 0.4G, comparable to the take-off on a modern airliner. Accelerating and decelerating to a maximum of 0.4G to attain 1000 km/h will thus take tens of kilometers. Moreover, for the trip to remain comfortable at high speed, limits need to be imposed on how changes in direction are conducted. This means that a hyperloop vehicle at high speed should not follow tight curves, which implies that hyperloop routes should be as straight and level as possible. The need to go as straight as possible and turn as gently as possible (beyond implications for curved tube and track assembly design) requires careful consideration in the selection of routes and securing rights of way.

In addition, hyperloop systems must include life support and other safety components, redundancies, and security and smart control systems. Although proven solutions from the aerospace industry can be adopted and extended to address some of these challenges, they can't be left aside as they are crucial to the system's overall viability.

Altogether, the dictates of the laws of physics make the design of hyperloop both challenging and exciting. Creative engineering and compromises are needed to achieve functional hyperloop system design. Some of these challenges are addressed in more detail in the following sections, along with some of the most viable solutions designers are currently pursuing.

The Evacuated Tube

Pressure, air resistance, and the Kantrowitz Limit. Atmospheric drag, or air resistance, is proportional to the square of speed. Double the speed and the resistance is 4 times greater. At 200 mph a racecar travelling in the Indy 500 faces 16 times more resistance than at 50 mph, and 100 times more than when it travels at 20 mph. Hyperloop addresses this challenge, in part, by removing most of the air from the hyperloop tube, thereby reducing the drag. The Alpha paper proposes a tube pressure of

100 Pascal (about 1 millibar), which implies removing 99.9% of the air. This equals the air density at about 50 km (30 miles) above the earth's surface, which is 4 times higher than the height at which commercial jets fly.

However, a hyperloop vehicle in a tube at 100 Pascal air pressure will experience a resistance greater than that for an airplane in flight at a height of 50 km (if it could fly that high). This is due to the fact that the limited space between the vehicle and the tube constricts the air flow of the air being pushed out of the way by the vehicle. As the vehicle's speed increases to a large fraction of the speed of sound (which is about 760 mph or 1225 km/h) the constriction of the flow becomes a limiting factor. The flow around the vehicle at supersonic speeds (that is, greater than 1225 km/h or Mach 1, the speed of sound) places limits on speed, which will depend on the ratio of the cross-sectional area of the vehicle and that of the tube, known as the blockage ratio. The speed at which air simply cannot flow around the vehicle fast enough, that is, when the flow is "choked", is known as the Kantrowitz limit. Interestingly, it does not depend on air pressure, although the pressure in the tube does influence how much energy is needed to push the vehicle to these high speeds.

The Kantrowitz limit significantly constrains a hyperloop vehicle, have on the tube configuration. For simplicity, visualize a vehicle with a circular cross section traveling in a tube, and consider the ratio of the diameter of the vehicle and that of the tube (the blockage ratio). A rough calculation reveals that, for a capsule moving at 500 km/h, staying below the limit requires a tube diameter of at least twice that of the capsule., or a blockage ratio of 0.5. Going faster, for example at 1000 km/h, requires a tube diameter five times larger than the vehicle, or a blockage ratio of 0.2.

None of the publicized hyperloop concepts propose a tube anywhere near five times the vehicle diameter. Materials cost, the extra weight borne by the structure, as well as the added energy needed to evacuate air from the large volume down to a pressure of 100 Pascal would make such an undertaking prohibitively expensive.

However, slight alterations in tube and capsule geometries do provide another way of relaxing the Kantrowitz limit. For example, bypasses (say between parallel tubes or even within a tube itself) could enable airflow to escape when pushed by the pod. In addition, perforated tube walls, common in high-speed rail tunnels, could be incorporated. Such walls have been demonstrated to effectively enlarge the cross-sectional area of a tube, thus decreasing the effective blockage ratio.

A hyperloop application designed exclusively for transporting freight instead of humans naturally opens new possibilities regarding speed and acceleration. Freight companies and their customers might be satisfied with relatively lower speeds compared to human transport, say, around 500 km/h, such that the Kantrowitz limit becomes much less critical. A speed of 500 km/h still represents an enormous improvement over railway and highway speeds and would be appealing to many companies.

The physics of a compressor. The Alpha paper proposes a clever solution to limitations on speed: use a compressor to actively move air from in front of the vehicle to behind it. A compression ratio of 20-to-1 is proposed, which means that the vehicle cross section could appear to be (in terms of corresponding fluid flow) 20 times smaller. That is equivalent to improving the blockage ratio by about 4.5 times. Using a compressor to achieve such a ratio, instead of requiring a tube diameter of 5 times that of the vehicle diameter to reach an absolute limiting speed of Mach 0.9, the tube diameter need only be a little over 2 times the vehicle diameter.

However, at the present, no existing compressor meets essential specifications for a hyperloop application. The closest comparable compressors are those used in modern jet engines. These do feature compression ratios of 20-to-1 or greater, but the inlet pressure for a jet, even at the highest altitudes flown, still greatly exceeds the 100 Pascal pressure desired in the tube. A hyperloop would thus require a very different design, which would take time to develop and test. Also, compressors require a lot of energy to achieve high compression ratios. Jet engines use jet fuel, which is not the anticipated power source for hyperloop. To meet power requirements for an electrically driven compressor, development and testing of batteries with higher energy density or new technologies to transfer power to the vehicle would be needed.

Finally, compressing a gas by a factor of 20 not only demands a great deal of energy, it increases temperature, creating the need to deal with the heat. A basic physics calculation concludes the compressed gas would reach temperatures of about 500 degrees Celsius, or over 900 degrees Fahrenheit. Although subsequent expansion of the gas when it is discharged reverses some of this temperature increase, a powerful compressor would need an additional system to dissipate that heat. This constitutes another complex engineering design problem. The Alpha paper suggests using a water-to-steam cycle where the heat eventually would be stored as water vapor in a highly pressurized steam reservoir. The full reservoirs are exchanged for empty ones at hyperloop stations. Although the physics behind this idea is sound, a great of engineering and perhaps development of new materials would be required to make it a reality.

Hyperloop companies could pursue new concepts by optimizing variables such as speed, pressure, and blockage ratio in absence of a compressor. Notably, none of the proposed capsules so far unveiled by hyperloop companies use compressors, although it is difficult to say at this point that the solution is being discarded, given that some companies have not publicly revealed solutions.

A lighter gas? In addition to the compressor idea, another approach to meeting engineering challenges involves replacing air with helium to address the choking limit, an idea that has generated a few patent applications. As the second lightest element after hydrogen, helium is much less dense than air. To give an idea of helium's relatively lower drag compared to air,the speed of sound in helium is three times faster than in air. If a vehicle is traveling in a tube with twice its diameter with helium, the vehicle can travel at over 1,000 mph before approaching the Kantrowitz limit at 380 mph (615 km/h). Although hydrogen has also been suggested, its higher reactivity creates formidable challenges.

Yet helium also comes with serious challenges, including the complexity of its introduction, its limited supply, and its difficult storage and handling. Helium cannot simply be blown into a tube, though some process could certainly be devised to remove air and introduce enough helium into the tube. Also, current limitations on the worldwide supply of helium put it in high demand. Its inertness and low temperature properties have made helium vital for many applications, from medical imaging to manufacturing of liquid crystal displays and fiber optic cables, and even the pressurization of rocket fuel tanks. There have been helium shortages in 3 of the past 14 years, causing price spikes. Supply is thus somewhat unreliable. Nevertheless, even with such challenges, using helium-filled hyperloop tubes may be worth pursuing. More research and testing are needed to fully understand how simple or complex its use may be. Currently commercial hyperloop companies have yet to announce that they will develop a helium-based tube environment.

Maglev

In contrast to drawing on the concept of air bearings described in the Alpha paper, much effort has focused on magnetic levitation, or Maglev. High-speed train projects like the French TGV and Japanese Shinkansen, even though they are moving through the atmosphere, have been able to reach speeds of 200 mph (320 km/h). Technical innovations made this possible.

As the high-speed rail industry strove for faster travel, it worked to reduce or eliminate contact resistance with the track. Maglev-based high-speed rail operates in Shanghai, China, with a 30-mile route routinely operating since 2002 at about 270 mph. Another system, using different technology, and has been developed and as is being tested. Japan Rail is applying the same technology to a commercial route being constructed between Tokyo and Nagoya, named the Chuo Shinkansen. It is scheduled for completion in 2027. Maglev speeds faster than 300 mph are possible with both technologies, but at considerable capital cost and with very large energy requirements.

Maglev offers promising possibilities for hyperloop, yet its prospects cannot rely simply on placing existing maglev technology in an evacuated tube. To understand this, consider the two different technologies by which levitation is accomplished in current high-speed rail maglev trains, namely, electromagnetic suspension and electrodynamic suspension.

Electromagnetically suspended systems, such as those used by the Shanghai Maglev (and first developed in the German Transrapid system), are based on electromagnets placed on the train cars that attract a ferromagnetic stator mounted to the track. The train's electromagnetic structure wraps underneath the track such that a net attractive force pulls upward from below, lifting the train off the track. The system requires tight tolerances given that the force of magnetic attraction varies inversely with the square of distance, so minor changes in the gap between the car and the track (usually less than 20 mm) produce large force variations. The system is dynamically unstable and must be tightly controlled to maintain a balanced state during normal operation, thus requiring a sophisticated, fast-response feedback control system to keep the train cars at the required gap. Electromagnetic levitation enjoys the great advantage of having a very low magnetic drag and working at all speeds, even when the train stops at stations, which eliminates the need for wheels, axles, or suspensions in normal operation. Designing a system with essentially no moving parts also greatly simplifies maintenance, reducing wear and tear on parts.

Alternatively, electrodynamic suspension systems rely on repulsive forces between the magnetic fields generated on the track and on the train car. The system must create very strong magnetic fields. It has a dynamically stable configuration where any change in the gap between the train car and the track inherently creates a restoring force that brings the gap back to its stable separation, thus eliminating the need for a complex feedback control system. This is the levitation technology used in the Japanese SCMaglev system being tested at the Yamanashi track. SCMaglev uses superconducting coils to generate the magnetic field required to levitate the train. The advent of improved high temperature superconductors that can be cooled with liquid nitrogen has made the system more affordable, but it still requires a lot of power for operation and cooling plants. Electrodynamic suspension-based maglev also needs a complex track for its magnet configuration. Unlike the case for electromagnetic suspension, the reliance of electrodynamic suspension trains on induced magnetic fields means the train must be moving at fairly high speeds before the magnetic levitation becomes strong enough to lift the train car. At slow speeds, or when stopped at a station, a mechanical suspension system and wheels are still needed.

Another form of maglev, known as Inductrack, has also been under development in recent years. Inductrack was developed at Lawrence Livermore National Laboratory as a passive, fail-safe form of electrodynamic maglev. One leading hyperloop company, Hyperloop Transportation Technologies, announced that their system will use an Inductrack-based levitation system, and most other hyperloop companies seem to be exploring the use of similar passive levitation systems.

Inductrack is based on an arrangement of magnets called a Halbach array and a track that has passive (unpowered) reacting coils. A Halbach array is a special arrangement of permanent magnets that reinforces the magnetic field on one side of the array while reducing the field on the other side. Ongoing development has shown ways to implement it using track designs as simple as slabs of aluminum. This fully passive arrangement of electrodynamic suspension does not need powered electromagnets or cooled superconductors, so it is potentially very energy efficient. Inductrack does not achieve levitation until the vehicle is moving fast enough to develop sufficient induced magnetic fields in the track. Like other forms of electrodynamic maglev, it requires a suspension and wheels for slower speeds.

Inductrack looks like a good candidate for hyperloop, but there are still challenges in designing an inexpensive and efficient levitation system based upon it. For example, magnetic drag arising from the magnetic field arrangement creates resistance to moving along the track. Like the electrodynamic levitating force, magnetic drag depends on the vehicle's speed. Although the levitating force increases with speed, the drag force increases with speed as well up to a point, until it decreases to a constant value. Managing the system design and operation with consideration of both forces is technically possible, but any such design requires fine-tuning and testing. Interactions between the levitation and propulsion subsystems of hyperloop, discussed next, make the challenge still more complex. A complicated feedback and compensation system likely would be needed to make the entire system - levitation and propulsion - most efficient.

Propulsion

Hyperloop systems require a propulsion system appropriate for a levitated vehicle that is not in contact with the track needs to be developed and implemented for hyperloop. Maglev trains rely on linear motors for propulsion, and these are likely to be the best solution for hyperloop systems as well. However, the design and implementation will certainly be different for hyperloop because of its tube environment, speed, and track configuration.

Linear motors can be seen apply an "unwrapped" configuration of traditional rotating motors. In a traditional motor the stator is stationary and the rotor rotates around it. In linear motors, the stator is laid out flat and the rotor moves past it in a straight line. The stator and rotor of a linear motor are sometimes called the primary and secondary, respectively. Two types of linear motors have been used as propulsion systems for maglev trains, linear induction motors and linear synchronous motors.

Linear induction motor. In linear-induction motors, polyphase (typically 3-phase) electric power is supplied to the motor windings, usually laid into slots of a straight laminated magnetic core. The core and windings constitute the motor's primary. Cycling the polarity of the phases generates a traveling magnetic flux wave that moves along the length of the primary. This magnetic flux wave induces eddy currents in the secondary (hence the name induction motor). The secondary typically consists of a conducting metal sheet, often with a magnetic backing plate. The induced currents in the secondary create an opposing magnetic field, repelling the flux wave and producing a linear force or thrust. The thrust created is very sensitive to the distance (gap) between the primary and secondary, decreasing as the square of the gap distance. Since one is on the vehicle and the other is on the track or guideway, the propulsion of such a linear induction motor operation is highly sensitive to the relative position of the vehicle to the track. It requires very stringent tolerances and control of the gap between the primary and secondary. This gap may be different than the gap for levitation. The propulsion track-based components sometimes are located to the side of the vehicle rather than beneath it.

Linear synchronous motor. In linear synchronous motors the primary still consists of a core and windings, but rather than the secondary having an induced magnetic field, it is made up of an array of magnets with a constant field strength. While small motors typically feature electromagnets, high-power applications like maglev train propulsion require superconducting magnets. In maglev trains, the same magnets are used for levitation as for propulsion. They are also the source of the magnetic field for the secondary of the synchronous motor. The linear synchronous motor produces thrust by locking the field from this secondary to the flux wave produced by the primary. This reduces the need for strict tolerances in the relative position of the primary and secondary. Furthermore, the locking operation of this type of motor results in higher power efficiency compared to the linear-induction motor. However, this efficiency comes at the cost of a more complicated implementation. The phased supply currents to the primary need to be precisely synchronized with the position of the secondary as the primary and secondary move relative to each other. The system must have extremely reliable position and speed sensing and sophisticated control electronics.

Both induction and synchronous linear motor designs require the primary to be powered. Linear-induction motors can have a fairly compact primary, suitable for installation in the moving vehicle. On the other hand, the intrinsic design of linear synchronous motors favors operation with a long primary that can only be accommodated in the guideway. This key difference makes linear-induction motors easier to implement for high-speed rail or for hyperloop, and with considerably less infrastructure cost since the track can be less complex. However, it adds the complications associated with locating the power source onboard the train or vehicle.

A maglev train in Nagoya, Japan, the Linimo maglev, uses a linear induction motor design with a short primary placed on the train cars. This impressive system opened nearly 15 years ago as part of Expo 2005. The Linimo maglev covers a 9 km route and reaches a maximum speed of 100 km/h, which is modest compared with the much longer and faster Shanghai maglev and Chao Shinkansen. Because of its short distance and slower speed, the Linimo maglev can be powered with onboard batteries.

So far, every high-speed, longer-distance maglev design, whether a prototype or a line in commercial use, including the Shanghai and the Shinkansen designs, relies on a linear-synchronous motor. This stems from another challenging problem: the need to generate, store, or transfer large amounts of power onboard a fast-moving train. Batteries, at least with current technology, would need to be large and very heavy to store enough energy to power a long-distance, high-speed train.

With a linear-synchronous motor, building the energized primary as part of the track has several major advantages. In this configuration the power to operate the propulsion plant can come from stationary power sources along the route, eliminating the need for a heavy power plant on board the vehicle or the need to transfer a large amount of energy to the vehicle while in motion. The vehicle can be lighter since it does not include the considerable weight of a primary propulsion system, and a lighter vehicle can accelerate and decelerate more efficiently. The linear-synchronous motor's placement of the primary in the track also has the advantage of making the on-board levitation magnets available for use as part of the reaction field (secondary), which results in more compact on-board components and higher system integration.

Unfortunately, this design also has a major disadvantage: linear synchronous motors, as used in maglev trains, are complicated pieces of engineering. Integrating the primary into hundreds of miles of track is a major and costly undertaking. A hyperloop linear-synchronous motor requires the installation of the primary stator (coils and/or laminated cores) as part of the track. Because continuously energizing miles and miles of a track is not feasible. power instead would be applied independently to segments of the track. That power needs to be precisely controlled since its frequency, phase and amplitude must exactly match the moving vehicle so the magnetic fields are properly synchronized. Therefore, each segment needs a power converter station, precise sensing equipment, and fast, reliable data communication to a controller.

Although linear-induction motors do not need such complicated track, as noted above, they face the complication of having an on-board power source. For the linear induction design, the onboard system must generate or store the same amount of power as would be supplied to the track-based propulsion. With present technology, storing that large amount of energy in the vehicle is nearly impossible, and considerable development would be needed to create a way to transfer the energy while the vehicle is moving at high speeds. Further, the same power converters and variable frequency generation are needed onboard, adding weight and lowering the efficiency and performance of the vehicle.

Both forms of linear motors present challenges. Although the major hyperloop developers have not announced how they will address these, one of the original concepts of the Alpha paper provides direction. If the problems of the Kantrowitz limit and drag described above can be addressed, then the hyperloop system can be made very energy efficient. The hyperloop could be powered by a combination of the two linear motor types. When accelerating out of station, sections of track could be installed for a high-power linear synchronous motor. An onboard linear-induction motor then could be used to maintain speed between stations. The onboard system would need less stored power if it does not provide the initial acceleration. Further, the onboard battery could perhaps be recharged through regenerative breaking.

An actual hyperloop propulsion system likely will be more complicated than that described above. The system design will need to account for many situations, including, for example, where speed or power is lost in the middle of the journey. The capsule must be able to continue to a station on battery power alone. Alternatively, if the onboard battery becomes depleted (again, for whatever reason), the capsule needs another way to proceed. Regulators may include safety rules similar to the aviation industry's requirement for "extended operations" (planning for the ability to reach alternate airports if weather or engine problems require it) to ensure a capsule can reach the next station even under a set of unlikely and unusual conditions.

Energy

Hyperloop holds the promise of being able to travel fast and far while using energy very efficiently. Companies have also said power for it will be supplied using clean, renewable energy sources. If challenges such as the Kantrowitz limit can be addressed, then moving vehicles very quickly with a relatively low energy consumption through hyperloop is feasible. However, other energy considerations in a hyperloop design must be considered.

From the point of view of physics and engineering, a hyperloop system design should reflect an understanding of all the energy needs associated with it, and over all the years of its life cycle. Even after having created a very low energy levitation and propulsion configuration, creating a vacuum in the hyperloop tube will be energy intensive (at least with current technology). The amount of energy needed by vacuum pumps to create a low-pressure environment grows exponentially as the tube pressure falls. Assuming a fairly common vacuum pump model, one calculation estimates it will take approximately one Megawatt of power to bring a fairly modest 8-foot diameter, 120 miles of tube length down to the hyperloop pressure of 100 Pa, and even this would take 35 pumps working for 5 days to achieve. An 8-foot diameter is actually relatively small for hyperloop considering that some company test tracks have already selected tubes that are 10, or even 13 feet in diameter. Though large, a megawatt of power may be acceptable if the tube can be evacuated once and then left sealed. The amount of energy needed for ongoing pumping to maintain that low pressure over the lifetime of a hyperloop line remains to be seen.

Although engineers have had experience creating low pressure chambers for industrial uses and scientific experiments, this does not come close to the large evacuated volume needed by hyperloop. The number of (presumably welded) joints and vacuum seals for the long tubes will be unprecedented, as will the number of airlocks needed to allow vehicles to be loaded and unloaded at stations while protecting the tube from outside pressure. Past experience with large vacuum chambers shows that joints leak, and low internal pressure must be maintained through constant pumping. The Large Hadron Collider particle accelerator, which is a 17-mile-long ring-shaped tube located underground crossing the border between Switzerland and France, has the largest vacuum system in the world. Its vacuum level is very different than hyperloop, but its volume is equivalent to only about 2 miles of an 8-foot diameter hyperloop tube.

Energy considerations in hyperloop tube design depend heavily on the target pressure. Lower pressure designs require higher energy for pumping, but lower pressure also lessens the energy needed to achieve high speed propulsion. Finding a pressure that balances these competing factors and optimizes energy efficiency will depend on many factors. This is another case where many of the hyperloop design variables interact with one another. The tube diameter and pressure, vehicle speed, weight, and carrying capacity, control systems, and much more all must be included as subsystems in a full systems approach to hyperloop design.

Control and Complexity

The overall complexity of controlling hyperloop poses an engineering challenge that will also take development. Think for a moment on the many facets that must be controlled and coordinated with extreme precision. The speeds involved, the sensitivity of the maglev and propulsion systems to the vehicle position along the track and its gap with the track, the need to minimize any air leakage into the tube, the need to optimize energy use, and the need to maintain passenger comfort and schedules all demand precision.

The details of the control mechanism must be handled by software and machines because humans cannot act and react at hyperloop speeds. The control and decisions will be largely autonomous. Some functions, like controlling the maglev gap and synchronizing propulsion between the vehicle and the track, will be done with on-board systems with strict feedback loops. Parameters must be kept within a narrow specification as the vehicle goes around curves and changes speed, and as parts age.

Other functions must be coordinated throughout a large network rather than just for a given vehicle. The vision for hyperloop transportation may be operating at aircraft speeds, but this must occur with multiple vehicles using the same track and departing every few minutes. In this sense, the system is more like a subway, except instead of vehicles stopping at each station, they go more directly to their destination.

Many examples already exist of complex onboard and network-wide computer control. For example, in aviation, military jets receive input from the pilot, but autonomously adjust the shape of the wing and aircraft thrust dozens or even hundreds of times each second. Conventional high-speed train networks also use complex control systems, both to maintain their network schedules and to control signals and switches along the track. Autonomous hyperloop network control must adjust the speed of vehicles to ensure continuous flow and avoid congestion. With so many requirements to control and coordinate hyperloop vehicles, tracks, and stations, hyperloop demands more complex and sophisticated control technology than has been envisioned for any previous transportation system. Moreover, hyperloop developers, together with regulators, must agree on interoperability standards so that hyperloop lines can seamlessly connect. The industry has yet to create such standards.

On top of the core issue of technical complexity is the overriding issue of safety. Autonomous control systems must be capable of reacting to any unusual situation that may arise. This could be a sensor failure, a problem or delay in another part of the system, an adjustment to schedules because of unexpected usage or demand, or more serious situations, like an unexpected change in tube pressure. Chapter 8 describes possible ways to thoroughly address safety issues for hyperloop with the aim of making hyperloop the safest mode of transportation ever created.

Summing Up

This chapter described many challenges that hyperloop developers face, and only some of the possible design directions to address them. These challenges should not serve as a deterrent to pursuing hyperloop today and advancing designs as quickly as possible. Sophisticated technologies continue to advance at an amazing pace. For example, just a few years ago autonomous vehicles were at a very primitive stage of development, but today they are about to share public roads with all drivers. Just a few years ago landing a rocket on a barge floating at sea seemed impossible, but now is done regularly. The list of impressive technical innovations goes on – human-like robots, hoverboards, 3D printing, and more. In light of this, meeting and addressing Hyperloop's challenges is not a question of "if," but of "when." Hopefully finding technical solutions will be accelerated by adoption of a spirit of open science and collaboration.

Rendering of an elevated hyperloop line in Thailand.

CHAPTER 4. Passenger Travel – The Hyperloop Pod Experience

Primary Authors: Mary Ann Ottinger and Thierry Boitier

People are eager to know what it will be like to ride in the hyperloop. How comfortable will it be? How spacious or confining? What will the acceleration feel like? What amenities can we expect? What human factors need to be considered in its design? This chapter discusses some design considerations specifically for passenger travel (which differ from those for freight).

Quite a few elegant designs of hyperloop pods and their interior layout have been created, and continue to evolve. They range in detail from artists' concepts to detailed engineering plans. Several companies are creating mockups of pod interior layouts, while others are focused more specifically on the passenger experience, comfort, pod atmosphere, simulated windows, and entertainment. University student teams and others are also contributing design ideas. Scale models and full-size pods have been built that give a visceral sense that hyperloop transportation will soon be realized. The elements envisioned for the pod interior will depend on the pod size and projected capacity. Seating design, flooring, layout, and facilities are all important for passenger safety and comfort.

Designers must understand how to prevent the anxiety passengers may have when they are travelling quite fast in a confined space, experiencing acceleration and deceleration, without a direct view of the outside. Thus, factors related to having a sense of time, place, and pace will also be key to the passenger experience. This includes acceleration and the perception of movement, in addition to the opportunity to move around, access onboard amenities, and view trip progress. This chapter uses just a few of the proposed pod examples to illustrate important considerations for human travelers.

In general, designers should aim to make hyperloop travel a wonderful experience – much more than simply a fast way to get from here to there. This will take strong attention to passenger comfort and ways of counteracting sources of anxiety.

Floorplan

In essentially all hyperloop pod designs, passengers occupy the central portion of the pod (see, for example, the figure below). The size of the pod is perhaps the first key design consideration. Since size impacts the availability of space for seating and amenities, bigger is probably better for travelers. But larger pods also have obvious drawbacks in terms of cost and technology challenges. As described in Chapter 3, the large-diameter tube needed for a large pod costs much more than a smaller tube. Also, a heavier pod needs stronger magnetic levitation, and this too adds expense and increases the energy needed for propulsion and braking. On the other hand, people and freight shippers may be willing to pay higher fares for a larger pod with a better passenger experience. Most current pod concepts are large enough to comfortably transport around 20 to 40 passengers, and companies in the industry are showing designs with a tube diameter between 3.3 and 4 meters (approximately 11 to 13 feet), comparable to the larger "Passenger Plus" hyperloop tube envisioned in the Alpha paper.

Seating arrangement concept, with screens in place of windows for passenger 'viewing' and information.

Seating

In most concepts that have been publicly shared, seating within the pod is configured in rows, as with planes, trains, and buses. The seats are central and concentrated, and there may also be room for restrooms and areas to walk. The next figure (as well as the previous one) shows examples of seating arrangements in rows within the pod.

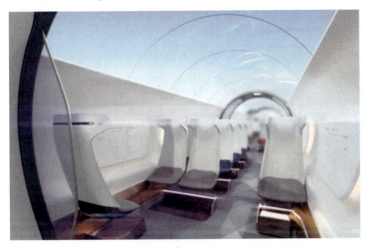

Potential configuration of seats arranged in rows within the main portion of the pod.

There are a number of potential configurations, including those that provide workspaces for occupants, perhaps even areas in which seat arrangements could be configured to provide time for discussion and meetings.

Designers also give close attention to the seats themselves. These must maximize passenger support, safety, and comfort during all phases of the trip. Typical acceleration in the hyperloop will likely be similar to the take-off stage of a commercial aircraft, with a forward acceleration of about 0.4g, but lasting longer as the pod reaches its glide speed, which is expected to be very smooth, with no further acceleration. Deceleration forces may be similar to the acceleration phase, although designers could choose a lower stopping force over a longer period as energy is recaptured through braking.

Besides these forward and backward forces, passengers need to be comfortable during the side-directed acceleration as the pod goes through curves. Seats and perhaps seat restraints must be designed to be comfortable and safe throughout. Pods may be designed with fixed seats that would always face forward whether the vehicle is accelerating or decelerating. However, some designs consider using banked turns, similar to aircraft coordinated turns, given that the sideways force becomes rearward or downward for passengers, although this could add considerable complexity to the maglev lift and propulsion.

These interior innovations are important for passenger comfort, but at high speeds any perturbation of the suspension systems also may lead to vibrations and motion sickness. The smoothness of the tracks and magnets, as well as the response time of the magnetic field to bumps and track irregularities, become critical considerations to reduce the potential for motion sickness and increase comfort and the passenger experience.

The willingness of passengers to use hyperloop for rapid transportation will depend in part on the arrangement and comfort of the seats within the pod. To address this key factor, hyperloop designers have the opportunity to build on the research and experience related to seat design from other forms of transportation. For example, NASA, military aircraft, and race car designers all have met seating safety and support challenges that are much more stringent than what is needed for hyperloop, providing that the technology can be transferred.

Lighting and Amenities

Lighting is critical to provide a sense of openness and to help settle passengers into their environment for the duration of the trip. Although overnight duration travel times, common for some transoceanic airplane trips, will not be needed for hyperloop, the passenger can still benefit from lighting adjustments to the time of day. For example, lighting might be made bright or soft, be dimmed in the evening, or be individually controlled by the passengers. Access to food, restrooms, and amenities like phone and internet have not been discussed much by the designers, but can also be very important to passengers.

Most plans also do not clearly specify the ability to move about the pod. As in an aircraft, passengers would not be able to move freely during acceleration and deceleration, and in hyperloop perhaps also not when travelling in curves. While subways, maglev, and traditional trains do not have seat belts, hyperloop, like airplanes and cars does require passenger restraints to suit its target rate of acceleration and deceleration. But during periods of glide or cruise, the freedom to move about becomes possible, and this makes a great difference in passenger comfort. Further, configuring pods with private sections for families to gather or for business meetings during travel seems attractive because people would be able to use their travel time for work or fun.

Finally, since travel patterns, electronic media, communications, and expectations will change given the rapid changes in technology and lifestyle that are sure to arise over time, the design of pod amenities should adopt a long-range view, considering life over the next 50 years or so. Thus, instead of focusing on creating something akin to a contemporary train or bus interior, hyperloop designers might ask themselves "how would passengers like to travel in 2030, 2040, or 2050?" and anticipate future trends as much as possible.

Interior Space. Avoiding a Sense of Confinement

Humans are uncomfortable when they feel confined in a small space, and they are more at ease when they have sensory input about their environment. In a pod with no windows traveling through a solid tube, visual and other characteristics of the physical environment become especially important.

Well-lighted pod interiors with an open-feeling and a sense of the outside (whether real or simulated) are critical to providing a feeling of calm and comfort. To relieve the perception of a small space, videos could play on the walls of the pod, or even a real-time outside view could appear on "windows" and in the front and back of the pod. Several companies have proposed this type of interior layout for the pod – turning it into a light and airy space that is peaceful and relaxing. Similar to an approach used on some AMTRAK trains, this pod concept allows riders to have great individual control over their local space and amenities (for example, choose which video displays) and tailor their environment to make use of travel time to get work done.

Since clean air and good ventilation matter for health, air quality must be given priority in pod design. Hyperloop pods must have air that is fresh, clean, and at a comfortable temperature. Scented air has been considered as another environmental enhancement. However, a 2019 study in the Vienna subway tested four different scents and showed that their passengers preferred no added scent.

Sensory Experience and Perceptions

As noted above, designers must account for the interaction of gravity and acceleration with human senses in the pod environment. Aerospace engineers and NASA space scientists have studied the human impacts of gravity, speed, and acceleration extensively. They have described how the human brain brings together information from the vestibular system (inner ear), visual apparatus (eyes), and balance centers in the central nervous system during travel. These issues are clearly important for astronauts and pilots; similar physiological responses to rapid acceleration and deceleration may also be experienced by passengers riding in the hyperloop pod.

Visual cues that do not match motion and acceleration detected by the inner ear can cause cognitive and physiological discomfort. Similarly, mismatches between aural cures and the expected environment may engender similar reactions. Because pods are designed to move in a medium-vacuum environment, passengers won't hear thermal engines, or the noise from air friction. They will hear background noise, such as electronics, conversations, etc. This may be very disconcerting to some passengers and contribute to the disorientation of passengers during the trip. Moreover, dissonance between these inputs could lead to nausea, vertigo, and problems with balance. Besides being very uncomfortable, these physiological responses can lead to an increase in blood pressure and an altered heart rate.

Hyperloop design can prevent sensory dissonance by using strategies developed for space travelers and the elderly. Providing visual cues that orient passengers can help in particular to neutralize conflicting sensory cues. For example, videos shown within the pod could be coordinated with actual changes in pod speed and direction. Although these cues are artificial, they may provide an effective orientation for passengers and real time feedback about the surrounding outside environment. In addition, the visual information can reduce the sense of confinement and its associated anxiety. Providing information, such as time to destination and other milestones, also will reassure those individuals who like to know where they are on their journey.

Magnetic Fields

Hyperloop needs strong magnets both for levitation and propulsion. Design engineers must ensure that electromagnetic fields encountered in passenger and freight areas will not present a hazard. Fortunately, there are international guidelines for exposure to magnetic fields (for continuous exposure, safe levels are approximately 1000 times stronger than the Earth's magnetic field). While inside the pod, passengers are likely to be well protected because the vehicle will act as a Faraday cage, i.e., an enclosure which acts as a shield that blocks electromagnetic fields. Nevertheless, field strength and the duration of exposure must be considered in other situations, such as when entering and exiting pods, waiting in a station, and during maintenance. With proper attention to system design, exposure to magnetic fields will not be an issue.

Summary

While many questions remain about the final designs and passenger response to riding in hyperloop pods, creative and detailed concepts have been proposed. Engineering, biological, architectural, regulatory, and business considerations will combine to create the most feasible, effective, safe and comfortable pod designs for human passengers.

CHAPTER 5. Infrastructure and Rights-of-way

Primary Author: Brad Swartzwelter

Hyperloop's success depends critically upon acquiring the right-of-way, i.e., access to the land and routes where hyperloop will be built. As usual and appropriate when creating a public good, governments must have a central role in this endeavor. Of course, once rights-of-way are secured, actually building the hyperloop infrastructure – the stations, tubes, and guideways – will be an engineering challenge. As typical for most transportation, both rights acquisition and the building of infrastructure to accomplish this great engineering feat require substantial funding.

It took a large investment to build our existing network of highways, railways, airports, shipping ports, and even neighborhood roads, all having considerable ongoing maintenance costs. After our societies invested in infrastructure for these modes of transport, they have provided an extraordinary return-on-investment that has grown our economies, wealth, and quality of life. The same can very well be true for hyperloop.

Smart design choices, including decision-making in light of long-term benefits, will pay off. Good infrastructure design for hyperloop will yield relatively small ongoing maintenance costs. Upfront route selection, sound hyperloop network integration with human and freight transportation hubs, and the establishment of straight, level rights-of-way for the network will allow it to be fast and efficient.

This chapter details the infrastructure needs and rights-of-way acquisition for hyperloop.

Land

Land is the foundation of everything people do on earth. A precious and finite commodity, land is often expensive to take possession of, even in small amounts. Any project humans attempt to build begins with land. A region's topography, hydrology, stability and density determine nearly everything else associated with a project – construction materials, techniques, appearance, cost, etc. When trying to create a land-based transportation system, land is extremely difficult to acquire because an uninterrupted ribbon of real estate must be secured before the project can go forward. As cities and suburbs have expanded, and open space has become scarce, access to the appropriate land has become a high hurdle for many transportation projects. Cities and regions building outer loop roads and beltways have this challenge, as do new airports built ever farther from city centers, and ports, which are unable to expand to adjacent land.

The great speeds that are characteristic of hyperloop make land requirements even more stringent than for roads or railways. Travel at a speed faster than that of a jet aircraft means hyperloop must be designed straighter than any long-distance land-based transport system ever built. Tight curves need speed limits, and the sharper the curve, the slower the pods must travel. Curves kill speed. The goal of hyperloop routes should be to have only gentle, large radius turns that will not slow the pod significantly.

As any good realtor knows, "location is everything," and this is no less true for hyperloop. Stations and freight facilities need to be close to where people are located and close to intermodal facilities, such as ports and airports. In part, that means creating new rights-of-way directly into the heart of cities – the most expensive real estate on Earth. Keeping the all-important continuous straight path through fully developed, owned and occupied cities will be challenging to say the least.

There are four basic ways land can be acquired for hyperloop.

1) Using eminent domain, a government can simply seize the land and move those who are already there out of the way. That works well in places nobody lives, like the desserts of Nevada or the Middle East. It's also a viable option in countries with strong centralized power. It does not work in densely populated democracies. Politicians that displace people from their homes and land tend to face strong

opposition, and therefore they usually reject calls for applying eminent domain.

2) Land can be bought at market value. A clean and simple way to get needed land, it is nonetheless nearly impossible to acquire each parcel needed to complete a project by buying at market value. A single hold-out owner that refuses to sell can make a huge project untenable. Also, since the route alignment requirement leaves few alternatives, the value of the land for a proposed route will skyrocket. In fact, as has happened with many projects, the prospect of a hyperloop line alone likely will cause land speculation and the cost of the needed rights-of-way may well become prohibitively expensive on the open market.

3) For some transport corridors, existing rights-of-way will serve directly for hyperloop because transportation stakeholders – in most cases governments – already own the continuous ribbon of land. Although not perfectly straight, roads and rail lines are nevertheless rather direct, and for some routes, they could be straight enough. However, in some cases the existing infrastructure will be considered the primary user of the right-of-way, and hyperloop will have to work around it (road, rail line, power line, pipeline). This potential competition between those with vested interests in the existing use and the proposed hyperloop may be worked out for some hyperloop routes.

4) Creating new land is not out of the question. Considered in three dimensions, humans are literally just scratching the Earth's surface. Every meter of tunnel and subsurface space chiseled out of the earth creates new land for human activity. Below ground nearly unlimited space exists for expansion and growth. Hyperloop can be as many lanes wide as needed underground, and have as many levels of track in as many directions as can be imagined. Two issues remain with digging lots of very long tunnels – time and money. Building tunnels and subterranean structures takes lots of time and expense. Important legal questions also arise about who owns land deep beneath the surface. Resolving such legal issues may be the first step to solve the right-of-way problem for hyperloop, and for a host of other related problems. In

the long term, going underground seems the most likely choice to establish an enduring network of hyperloop routes.

Returning to the geometrical requirements for hyperloop, not only do curves need to be relatively gentle for hyperloop, slopes must be as well. Consider hyperloop crossing one of the world's great mountain ranges. No straight-line path across this terrain exists. Surface routes for roads and power lines tend to follow the easiest routes – through valleys, over passes, along rivers. In some cases, such surface routes may suffice for hyperloop. But for a straight, flat right of way, tunneling is always available to meet the requirements.

Where high-speed rail and highways require a 50-yard-wide ribbon of land, often cutting properties in half and rerouting existing roads and utilities, hyperloop tubes can be 20 feet above ground on pillars. In this case existing infrastructure is less likely disturbed, and agricultural machinery will be able to operate below the tubes. Alternatively, the land required for each pillar could be leased, similar to agreements for windmills.

Tubes – High or Low?

The question of right-of-way is linked to the decision to build hyperloop infrastructure primarily at surface level, elevated on pylons, or in tunnels below ground. Because of the many obstacles, even on relatively low terrain, building along the ground is probably less likely than constructing an elevated system. Of course, multiple methods can be combined. Japan Rail's Yamanachi high speed rail test line (not a hyperloop) combines long underground segments, short tunnels through hills, some segments at-grade level, and elevated tracks on pylons.

Above or At the Surface Level?

When the first hyperloop lines are built, they will probably be above ground due to speed of construction and lower cost. That means either using existing rights-of-way in developed areas, or acquiring land by whatever means possible. If, for example, existing interstate highways in the United States are used, the probability is that most infrastructure will be tubes elevated above ground with pillars placed between the lanes of traffic. Elevated tubes do not complicate highway interchanges.

Consider, as an example, the proposed route between Kansas City and St. Louis, Missouri. U.S. Interstate 70 is a very straight and flat highway for the 400 km (248 miles) between these cities. More than 30 meters (about 100 ft.) of open space predominates between the east and west lanes of traffic. Already controlled by the government, this land could not only accommodate the tube and underlying track, but also photovoltaic solar farms to power the system.

Current land transportation networks were built at surface level, but these were put in place when the U.S. population was smaller, and land was far more available. While building at surface level still offers the least expensive construction costs, better long-term options exist. For example, the early city subway line of New York City used shallow "cut-and-cover" trenches. Placing hyperloop similarly just below the surface has some advantages compared with surface or elevated tubes. The approach presents several benefits: infrastructure would be separated from any impacts with traffic; the materials would not be as affected by temperature fluctuations like thermal expansion and contraction; proximity to the surface allows a rapid escape system in emergencies; and minor undulations in the landscape can be smoothed over.

Yet cut-and-cover construction can be very disruptive in populated areas. The method is rarely used today, even in moderately populated regions. The city of Vienna Austria was able to use cut-and-cover to build parts of its wonderful subway system. But a major road in the city, Mariahilferstraße, representing just a few kilometers to the transit system in Vienna, remained a construction zone for nearly the entire decade of the 1980's. Area businesses suffered and people living near the construction zone were severely inconvenienced.

Construction of the 2ⁿᵈ Avenue subway line in New York City disrupts travel and commerce in 2010.

The impacts of weather must also be considered for above-ground hyperloop infrastructure. Long steel tubes should allow for thermal expansion of the metal, while supported by an unmoving foundation. Whether on pylons or at surface level, hyperloop infrastructure must be built to deal with a range of temperatures, temperature swings, wind and snow loading, and even occurrences like slow surface movement and occasional ground tremors. Engineers have studied similar issues, for example with long rail lines and pipelines. Hyperloop may need additional research, but the engineering design challenges can be solved.

Part of the vision for hyperloop right-of-way includes serving as a space for the placement of photovoltaic arrays. Designers estimate that that power from solar panels over this area can be more than enough to supply all aspects of hyperloop with carbon-free electricity. Carbon-free electricity represents an important resource, not only for hyperloop, but also for all of society.

Deeper Tunnels

Hyperloop needs structural stability and straight lines, and deep tunnels would allow both. At a depth of fifty meters people on the surface will not see, hear, feel, or sense the construction or operation of hyperloop in any way. During construction, tunnel boring machines could grind forward day and night with a precision measured in millimeters over distances of many miles. But perhaps the greatest advantages of tunnels are their durability, longevity, and low maintenance.

Tunnels last a very long time: Some of the oldest operational infrastructure on earth are tunnels. King Hezekiah from the constructed a water tunnel 2,700 years ago that still functions in Jerusalem to this day.

Tunnels are resilient: All modern modes of surface or air transportation can be destroyed by a single shot from a .50 caliber rifle at 500 meters range. Hyperloop in deep tunnels will remain functional even after an atmospheric blast of a nuclear weapon.

Tunnels are fast: Curves kill speed. Only tunnels can precede in perfectly straight lines over great distance regardless of development or topography.

Compared with above-ground options, time and money present the greatest challenges for creating tunnels for hyperloop. But technology for underground construction continues to advance, bringing down these costs. Modern tunnel boring machines are nearly autonomous, and in favorable, solid-rock conditions, they can advance as much as fifty meters a day, creating a brand new, laser-straight right of way. At the surface, the impacts of construction are limited to pits, where the machines start digging, and through which ground tunnel material is extracted and tunnel lining material is introduced. If a series of pits are created at fifty km intervals, a fleet of tunnel boring machines could create a continental scale hyperloop tunnel in just a few years – likely less than five. Further, the cost of using tunnel-boring machines has dropped dramatically in recent years, in part because of automation.

The cost of tunneling goes up as the size of the tunnel diameter increases. Current industry tubes are proposed at 3.3 to 4 meters. If a hyperloop will be transporting existing intermodal containers, which require substantial tunnel lining, tunnel boring machines will need to handle a much larger diameter, perhaps as large as 5-6 meters.

The modern tunneling process is fascinating. Behind the cutting head of the machine, sections of 30 cm-thick liner rings are bolted together. The rings are resin-coated, pre-cast concrete with rubber gaskets. Grouting injected between the rings and the earth micro adjusts the tunnel structure to be perfectly aligned. A waterproof, air-tight, highly-bondable membrane is applied to the inside of the rings, and an interior ring with the shape needed for hyperloop maglev gear is bolted into place. This tunnel lining process is known as the sandwich method. In most circumstances, tunnels built with materials available today should function for hundreds, and perhaps thousands, of years. The tunneling and lining process currently costs about US$40 million per km for twin tunnels through most ground types.

Tunnel boring machine at construction site of metro in Bangalore, India.

Stations (Terminals)

Hyperloop stations may well become the signature architectural achievements of the 21st century. Like concert halls, cathedrals, and museums, transportation hubs like major train stations and airport terminals reflect the way of life and culture during the period when they are built. Because hyperloop stations will be new and loaded with promise, communities will want to create designs that emphasize the unique characteristics of their region and that integrate into the locale. This was the case for the "tents" of Denver International Airport, as well as for the warehouse design of Oriole Park baseball stadium at Camden Yards. Both became icons for their cities and were a vanguard of similar efforts by other cities. Cities may compete to create grand hyperloop stations with natural lighting, open and breezy passageways, and easy interconnections to get to other parts of town. Although the first hyperloop stations will most likely be found in city centers, smaller and decentralized stations can be located in less concentrated areas.

Hyperloop terminals for freight are likely to have a more industrial architecture that is functional and efficient. Freight stations will probably be at surface level regardless of whether the main line tubes are elevated or in tunnels. The terminals will be located near ports and rail yards, and have automated flow of containers and pallets between various modes of transport. Integrating hyperloop with all other intermodal transportation will be vital over the decades it will take to build hyperloop. The system must work together with road, rail, ship and air freight throughout transition and beyond. Other terminals are likely to integrate directly with the shipping centers of large companies and package services (such as: national postal services, Walmart, Amazon, FedEx, DHL, and UPS).

Airlocks

Platforms, airlocks, and connections to other transportation will be defining features for all stations. Two kinds of airlocks have been proposed, and it's likely both will eventually be used. The first is a classic airlock for the full pod. When arriving at a station, the decelerated pod would move off the main tube and into a depressurized airlock, where, once sealed, the pressure in the airlock would be matched to outside atmospheric pressure so people or cargo could enter and leave the pod. Upon departure of the pod, the sealed airlock would be evacuated and the pod reintroduced to the hyperloop high-speed tube. The second type of airlock would limit pressurization to "bridge doors." The doors would form a seal around the outside of the pod doorway and act as a temporary pressurized passage for people to enter and leave. This would be like the jetway for loading and unloading aircraft, but with the added pressure seal. It is likely the first (full pod) airlock option will be used for initial hyperloop lines.

Hyperloop pods specialized for freight will surely have much different entrance and exits systems than those for passengers. The freight pod will need large bulkhead doors. Rollers might be used to facilitate the movement of intermodal containers being taken out of the pod and whisked away to the next mode of transport. Massive marshalling yards for freight may be needed at busy ports and city terminals. Early indications are that these could be built in remote, less populated locations. For example, while ports today are very crowded and occupy valuable coastal real estate, shipping containers could be quickly moved to an inland area where they would either continue in the hyperloop network to their destination or be distributed to other forms of transportation.

Safety

Safety infrastructure for hyperloop will be very similar regardless of whether the alignment is above or below ground. The greatest areas of concern are losing pressurization of the pod inside the low-pressure tube or crashing into something at 1000 km per hour. System design and a culture of safety should make these concerns very low probability. Situations involving risk may arise that are analogous with those encountered by a modern airliner. Yet air travel has a remarkable safety record, and hyperloop travel can be made even safer. Certainly, safety features must be integrated into the design to give safety redundancy and survivability to the system.

Earthquakes pose a risk for any ground infrastructure. In the case of hyperloop, a significant earthquake could not only misalign the tubes, disabling pod travel, but also create breaches in the tube, provoking a sudden and dangerous flow of air. We know how to build infrastructure that can resist earthquakes, but this comes at a major additional cost. Hyperloop lines most likely will not be able to be built in earthquake-sensitive areas.

The Future

Once hyperloop benefits begin to be realized, the network can expand exponentially around continents and around the globe, just as paved roads did in the last century. The transformation to fast, efficient, inexpensive, carbon-free, transportation can launch new communities, dramatically change property values, and perhaps bring the physical world together in the way the internet connected the world digitally.

This connectivity will not happen overnight. The first attempts at hyperloop prototypes indeed have been built, and the concept has been tested, but on the very small scale of test tubes less than one mile long. Longer test tracks, of perhaps ten to twenty km in length and maybe with a curve, will soon be constructed. These tests sit on short pylons, but in coming decades, as the network grows and lines penetrate city centers, hyperloop will probably be underground stretching for many kilometers. As hyperloop expands, and its value to travelers grows through the network effect, it may well become the largest structural engineering project in the history of the world.

CHAPTER 6. The Funding and Financing of a Hyperloop Network
Primary Author: R. Richard Geddes

Introduction

Despite widespread engineering study and operational hyperloop test tracks in several countries, the question of how to address the formidable set of economic and policy challenges posed by a hyperloop network has hardly been explored. These challenges include possible sources of funding to pay for the network, what types of financial structures and instruments would best finance it, how the necessary right of way would be acquired, how environmental impacts will be addressed, how various impacted stakeholder communities will be engaged and how the relevant governmental jurisdictions will cooperate, among many others.

With the assumption that the engineering and technical problems posed by hyperloop transportation can be addressed, and that a hyperloop network is technologically feasible, this chapter focuses specifically on addressing economic and policy issues.

The key distinction between *funding* and *financing* hyperloop infrastructure often causes confusion in infrastructure policy. Funding is generated from the gross revenues of operating hyperloop. These revenues are spent on operations, maintenance, and compensation to investors for their risk, for example through interest payments on debt. Financing refers to the use of financial instruments (such as tax-exempt municipal bonds, taxable corporate bonds, etc.) to generate the enormous up-front payments required to design and construct such large infrastructure facility before operations can begin. Critically for public policy, financing is not available from financial markets unless adequate funding is in place. Issuers of financial instruments, such as bonds and equity, will demand that their investments be paid back with compensation for risk. For sufficiently "bankable" infrastructure projects, however, vast pools of patient, global capital stand ready to provide the necessary financing, often at low rates. *Funding* is thus the key prerequisite for an economically successful project.

This chapter describes both funding and financing as they apply to the hyperloop. It describes the key insight that funding rather than financing serves as the most fundamental challenge. Hyperloop's future relies critically on demonstration of its economic feasibility, where user fees and other revenue sources can cover capital and operating costs without ongoing subsidies. This chapter also stresses the need to focus on increasing the market demand for hyperloop, largely by differentiating hyperloop from other modes of transport. Further, it discusses concepts from network economics that are applicable to the hyperloop, and describes how a hyperloop network could be usefully governed. When properly applied, network concepts can help increase the funding available from user fees.

Funding the Hyperloop Network

The long-term economic feasibility and sustainability of a hyperloop network depends primarily on the gross revenue it produces. Scale and learning-by-doing can help to manage and reduce costs over time, but revenues are determined largely by customers' willingness-to-pay for services. This willingness must be sufficient to generate enough revenues to cover the prodigious capital costs of the network, as well as ongoing operation and maintenance costs. Indeed, covering the high fixed costs of design and construction (often associated with debt service) are a distinguishing characteristic of much network infrastructure.

For any type of infrastructure, funding sources can usefully be divided into three broad categories, which are distinguished by how closely the funding source tracks payers' infrastructure usage. One category closely follows use via a user fee, such as a per-unit toll, rate, or price that is charged directly to the user. Utility service pricing provides a familiar example, where customers are charged per minute for cell-phone communications, per kilowatt hour for electricity, per gallon of water usage, and so on.

Funding may also be established through some type of dedicated or targeted tax revenue. Taxes levied by a relevant government can be dedicated to funding a particular infrastructure facility or sector (such as transportation). Many such approaches fall under the broad concept of *value capture*, where the installation of the new infrastructure creates added value. The tax captures some of the newly created value to pay for the infrastructure. Under the prominent value capture method known as *tax-increment financing* (TIF), a government entity (i.e. the infrastructure owner), often a municipality, defines a geographic area called the TIF district that is likely to benefit from the infrastructure's installation. The district typically extends well beyond the project site and into the broader region, where property values are expected to increase due to the new infrastructure. Some of that increased property value can be captured via property taxes. The municipality pre-commits the expected increased revenues to repaying debt (usually in the form of bonds), enabling it to borrow funds dedicated to designing and constructing the new infrastructure. For example, municipal bonds could supply funding for a new subway stop that is expected to increase nearby property values, and thus generate increased property taxes. Value capture relies heavily on financial markets to monetize the future new revenue source.

In addition to direct user fees and value capture, general tax revenues provide a third option for funding. General tax revenues are only weakly related to facility use, if at all. For example, a state could raise its sales tax and dedicate the added revenues to funding transportation infrastructure. Such a tax is not a price, rate, or user fee linked closely to project use in any sense, but simply a funding mechanism for the project.

The rest of this chapter focuses on user-fee-based funding, the likely source of the bulk of revenue for hyperloop and the main revenue source for most public utilities.

Funding a Hyperloop Via User Fees. The majority of public utilities are funded by some type of user fee - a rate, a charge, or a price. The term "price" refers to all mechanisms that charge for use. Some unit prices vary over time, while others are constant. A variable unit price may depend on current demand and supply conditions, such as at peak times of day or real-time vehicle flows. For example, a variable unit price might be applied to managed or high-occupancy toll lanes.

Variable, flexible, per-unit prices offer many benefits. Most importantly, they provide critical "price signals" for allocating scarce infrastructure resources to their highest-valued use. Price signals encourage customers to conserve a scarce resource by, for instance, changing the time of their trip or seeking other transport options. As applied to the hyperloop, such actions help ensure that the highest-valued uses of the tube get priority.

Price signals also serve the less-appreciated purpose of allocating or directing supply. In the short run, high prices encourage providers to make more of a resource available. In the longer run, high prices indicate where new capacity should be added or subtracted. Absent price signals, investment in new capacity or in the reduction of unused capacity becomes more exposed to politicization.

Applying these concepts to pricing hyperloop use, note that many public utilities slowly transitioned from rigid, regulated prices to prices that vary in real time. Hyperloop developers should think early on about such lessons from variable pricing applied in other utilities. Hyperloop could benefit enormously from creating a system of highly flexible prices from the outset. The initial structure of charging for hyperloop tube use should allow prices to change rapidly in response to current demand and supply conditions. That includes the creation of both spot and forward markets for hyperloop use.

The novelty of hyperloop pricing leaves many basic issues unresolved. What would be the most *appropriate unit* for charging customers? Should pricing be done by duration of using the system (such as per minute), by distance, by weight (such as the pounds moved per mile), or by some combination of the above? The appropriate unit may also vary for passengers versus goods.

Economics offers insights here. The preferred unit can be discovered through experience by a market-based trial-and-error process. At the outset alternative ways of charging for hyperloop use will be attempted. The best method will likely feature the lowest transaction costs, but those costs must also be discovered through the market process. Solutions that are higher cost or more cumbersome will be discarded.

Hyperloop's novelty presents the opportunity for pricing unburdened by historical practices and technology. New technologies allowing accurate, real-time use tracking and metering enable much more precise pricing than in the past. This, in turn, allows better and faster response to price signals.

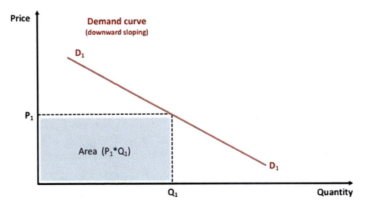

The Price-Quantity demand for hyperloop use, holding all other factors constant.

After resolving the basic issue of how to charge for hyperloop use, developers must next consider how to generate the greatest revenue possible from user fees to fund operating and capital costs. Economics here suggests turning to the most basic microeconomic relationship: the demand for hyperloop use. As shown in the figure below, the demand curve is represented by a downward-sloping curve in price/quantity space. The price willingly paid per unit by consumers (vertical axis) falls as customers use more units (quantity, horizontal axis), this relation known in economics as diminishing marginal utility.

The area under the curve ($P_1 * Q_1$) represents the amount of gross revenue generated per period, which varies depending on the point held on the demand curve. At too high a price, no customers will use the hyperloop. At too low a price, many will use hyperloop, but the revenue per use will be tiny. Given the existence of some price/quantity combination that maximizes revenue, price must be chosen carefully to achieve this point.

Gross revenue produced by hyperloop (which determines its economic viability) can be thought of in two ways. First, revenue can be viewed in terms of the size of the market, that is, how many units of hyperloop customers are willing to use at a certain price (how far the demand curve sits from the origin.). Alternatively, revenue can be understood in terms of how much customers are willing to pay for a certain level of hyperloop use. Both of these assume other factors, such as the income of customers and a constant value for the price of alternative options, are held constant.

An increase in the value of a hyperloop network can be understood in several ways. They can be represented via a rightward shift in the demand curve above, as shown in the figure below. The demand shift from D1 to D2 can be interpreted in two ways. Price can be held constant, while the quantity that customers will be willing to use at that price increases in the shift. Alternately, quantity of use could be held constant, yet customers are willing to pay more for that level of use. Intuitively, this reflects an increased *value* of a hyperloop network to customers.

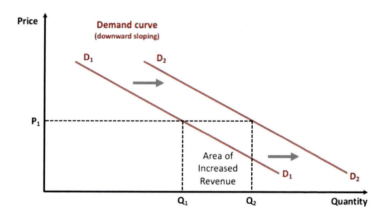

Increasing Demand (and thus Funding) for Hyperloop Use.

This figure also shows that the firm's revenue (and thus available funding) unambiguously grows as demand shifts rightward. Thus, to make the technology economically feasible, the firm should focus on increasing the willingness to pay for hyperloop use. The next section describes some ways firms realize more revenue by thinking ahead to increase value and shifting the demand curve outward, and thus enhance the economic feasibility of a hyperloop network. The greater the numbers associated with the demand curve shift, the greater the gross revenues available.

In addition to better understanding how increasing demand leads to increased value, these figures also lend insight into the *elasticity of demand*. Demand elasticity for hyperloop can be thought of as the price sensitivity of demand. It formally can be defined as the percent change in the quantity of hyperloop use demanded for every percent change in the price of its use, which is represented graphically as the slope of the demand curve. Higher elasticity of demand shows up as a flatter demand curve, where a relatively modest change in price leads to a relatively larger change in quantity. For example, a 10 percent increase in price will result in a larger than 10 percent fall in quantity (use, ridership). If the demand is *inelastic* (less price sensitive, generally corresponding to a steeper demand curve), then a 10 percent increase in price will result in a smaller than 10 percent fall in quantity, i.e., quantity (use or ridership) was less sensitive to the price change.

These figures make clear that a hyperloop firm prefers to face inelastic rather than elastic demand. In this case, the firm can raise prices without losing too many customers to other transport modes. All else being equal, the firm under inelastic demand can collect more revenue than under elastic demand.

Intuitively, this implies that a firm must strive to differentiate hyperloop transportation from other modes. Thus, the hyperloop transportation experience must be perceived as "unique," meaning (in economic terms) no close substitutes exist for hyperloop travel. In an environment where several other mature transportation modes are available, this becomes challenging. Yet hyperloop indeed has some very distinct points of differentiation. For example, once mature, its reliability should be excellent, uninterrupted by weather or congestion.

In summary, hyperloop providers and the industry must assign primary importance to focusing on the demand for its use, as well as differentiating hyperloop from transportation alternatives.

Strategies for Increasing Demand for Hyperloop Use. Although some come with significant costs, the value of a hyperloop transport system to customers can be enhanced in several ways. This section describes two strategies: commodification and network principles.

The value of something can be enhanced by standardizing or commodifying its use as quickly as possible, making it appealing in the short term to a large number of customers. Despite the often pejorative meaning assigned to commodification, as when people describe something as being reduced to a "mere commodity," for many industries, and particularly those involving a large network, commodification creates benefits. For example, the commodification of an airline seat on a specific route has facilitated fierce competition, and this has led virtually all airlines to have loyalty programs to resist such market forces. In so doing, airlines differentiate themselves in an attempt to make the demand curve for their particular service less elastic.

From an economic perspective, commodification of hyperloop reduces its transaction costs (roughly referring to the cost of using the market mechanism) and increases demand. Commodification may look quite different for passenger versus freight transportation, but can be achieved in both markets. Containerization of shipping increases demand for freight transport, among many other methods for achieving this.

Commodification importantly encourages the development of a large, fluid market for hyperloop use. With such a market, prices can then be set more accurately to counter congestion and allocate supply. A more fluid market also allows for the formation of both spot and forward markets in hyperloop use, something that should be done as soon as possible. Forward markets allow customers (particularly large customers) to hedge against the risk of future price changes. They also spur the emergence of "service providers" who offer services, such as hedging risk and monthly service plans.

In addition to leveraging commodification, demand can also be increased by exploiting hyperloop's clear potential for network benefits on the demand-side of the market, more precisely called for *positive network externalities*. This occurs when the purchase and use of a good by a greater number of users benefits all consumers, and contrasts with a *negative network externality*, where adding additional consumers imposes a cost (as opposed to a benefit) on other customers. Traffic congestion serves as a canonical example of such a negative network externality.

The familiar telephone or communications network helps to exemplify positive network externalities. Its value increases exponentially with the number of other customers on the network. As illustrated in the next figure, if there are only two customers on the network, a phone has minimal value: the two customers can only call each other. If the number of customers increases from two to five, however, each user can call four others on the network, increasing the total number of connections to 20. With 12 customers, the value increases even more, as each telephone has a higher value for each user. This helps explain the enormous demand for the modern cell phone.

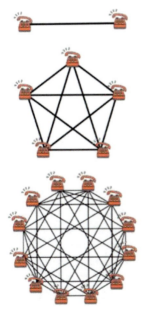

Network Value Increases Exponentially in the Number of Nodes.

Network industries in general characteristically exhibit such externalities. Considering airports in place of telephones, each airport's value is much higher when flight connections are plentiful rather than few. The value of a hub-and-spoke system in airports, then, comes as no surprise, not to mention that of the very node-rich internet.

With these concepts in mind, a hyperloop network will increase in value (again increasing demand and thus revenues) with an increase in the number of stations connecting hyperloop routes, as well as an increase in the number of routes to which each station connects.

Many networks, such as aviation and electricity networks, evolved organically, with little attention paid to the network effects that would eventually emerge. A hyperloop network could instead benefit at the outset from today's much greater understanding of how networks operate. Rather than focusing on individual routes, concentrating on a network of routes from the outset may well enhance economic feasibility.

Financing the Hyperloop Network

Financing for a hyperloop network can be secured once a steady stream of funding is anticipated from user fees or other sources. Financing provides the large upfront payments necessary to design and construct the network, and can take many forms, including tax-exempt municipal bonds, taxable corporate bonds, equity investment, and TIFIA (Transportation Infrastructure Finance and Innovation Act) loans, among others. The mix of financing tools will be tailored to the particular hyperloop project or route at hand.

Project financing will likely emerge as a financing approach. This form of financing relies on a non-recourse or limited-recourse financial structure, that is, the debt and equity used to finance the project are paid back from the cash flow generated by the project alone (in this case, from hyperloop user fees). The project's assets become the collateral backing the debt, a tactic that insulates participating parties from greater risk. Importantly, the definition of the "project" for financing is open to interpretation. In the hyperloop context, the project may be an entire route, or perhaps segments of a longer route that are divided up into the optimal units for financing and construction. The best projects are discovered and defined via the market process.

Project finance may also entail the use of a special purpose vehicle, or SPV, to deliver the project. An SPV serves as a separate legal entity created for the sole purpose of completing the project. It often carries the project name. The SPV raises the significant capital required for a project by issuing debt and equity. The various parties who must cooperate to deliver the project – design companies, construction companies, operating and maintenance companies, banks, among many others – enter into a contract via the SPV. The SPV in turn bids on the project, constructs the project, engages in operation and maintenance, etc. The figure below shows the SPV structure. In the hyperloop case, the "off-take purchaser" (normally an energy company) would be the buyer of hyperloop services, likely a hyperloop company. The SPV would be delivering hyperloop services for the purchaser.

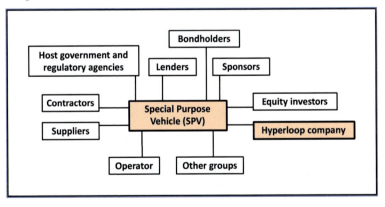

The Structure of a Special Purpose Entity under a Project Financing.

An SPV project financing structure is indispensable for managing the formidable risks inherent in many infrastructure projects. It effectively insulates the parent companies that created the SPV from project risk. Such a structure is critical to encourage large companies to undertake major infrastructure projects.

Equity investment in the SPV is particularly important. Equity investors have "skin in the game" in the sense that they bear residual risk – they are the project's residual claimants. Although their downside risk is limited to the amount they invest, their potential upside returns are unlimited. If the project does well, they capture the added returns.

Notably, different types of equity investors will participate in the project over its life cycle. For any greenfield infrastructure project, the earliest stages pose the greatest risks since the size and stability of project revenues are less clear. This higher level of risk is often referred to as market risk, commercial risk, or revenue risk. Investors willing to accept greater risks in exchange for higher expected returns typically hold project equity in the early stages. These may include specialized private equity funds, for example. Such investors often sell their stakes once project cash flows stabilize, and the project is de-risked, after a few years. At this stage, large, long-term institutional investors such as pension funds and insurance companies, often buy those equity stakes. Relying on such established financing approaches would aid hyperloop development.

The Cost of a Hyperloop Network: Scale Economies and Learning-by-Doing

To focus on economic value, the above analysis considered gross revenues only. Investors, however, are focused on net cash flows produced by the project, the difference between gross revenues and project costs. Costs include operating and maintenance costs, as well as contributions to capital costs (such as debt service). Costs are thus also important to consider. Indeed, in making the hyperloop economically viable, the complement to increasing gross revenues is reducing cost. For this analysis two key economic concepts related to costs stand out: economies of scale and learning-by-doing.

Economies of scale are also distinct from the network economies discussed above, and involve what is known as returns to scale, concept often confused with other economic effects. An economy of scale refers to the notion that the cost per unit falls, rendering cost savings, as the overall size, or scale, of the firm rises. For example, a water system requires a large fixed cost of installation before the first gallon can be delivered. If a water system initially was to serve only a few households, then the costs per gallon would be very high. As more and more households join the water network, the cost per gallon would fall because the fixed cost of installation would be spread out over many households. Scale economies are thus a very important effect for much networked infrastructure, which typically features high fixed costs.

An inherently long-run concept, returns to scale refers to the impact on returns (revenues) when the relation between production (output) to input changes at scale. Over an extended period of time, a firm can adjust all project inputs to the production process. Returns to scale differs from the impact on returns due to inputs that affect a fixed factor, such as adding more workers (input) to a fixed number of machines.

A hyperloop network is likely to exhibit large scale economies over a wide range of outputs, and as a result the cost per use (perhaps per ride) will fall as the number of users rises. In other words, if only a few riders were to use the hyperloop system, then the cost per use will be relatively high, but if the number of users rises, as expected, then the fixed costs of installation will be spread out over more users, reducing the cost per use.

In terms of policy implications, such expectations suggest that hyperloop developers should strive to increase the scale of the hyperloop network as quickly as possible. An extensive network, benefitting from the positive network effects discussed above, would attract many customers. That, in turn, lowers cost per unit, and thus the price charged to users, further enhancing hyperloop's appeal. Economies of scale are illustrated in the figure below. The curve can be viewed as the hyperloop firm's long-run average cost curve.

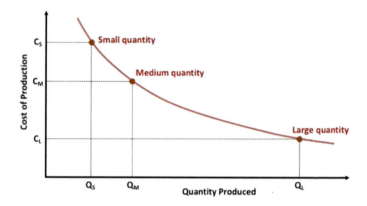

Economies of scale, where Quantity Produced refers to hyperloop use

Hyperloop costs can also be positively impacted by what is known as *learning-by-doing,* which refers to the idea that developers will get better at delivering as they gain more experience in design, construction, operation, and maintenance. The effect of learning-by-doing can be viewed as shifting the long-run average cost curve down over time. Given economies of scale, learning-by-doing reduces the cost of producing a given amount of hyperloop services.

Conclusions

The hyperloop represents an exciting new mode of transportation, the first in over 100 years. While engineering issues have received great attention, the critical questions of how a hyperloop would be funded and financed remain relatively unstudied. However, the economics and experience from transportation infrastructure projects provide strong guidance.

In short, the bulk of funding – the income generated by use of the hyperloop – should come from user fees in the form of real-time prices that change quickly in response to supply and demand conditions. Prices set in both spot and forward markets would allocate demand and reduce or eliminate the political direction of infrastructure spending, which has been endemic for hundreds of years.

Financing for the hyperloop should use the project financing model, through creation of a special-purpose vehicle. This would encourage key industry players to bear risks they would otherwise shun.

Hyperloop development can be accelerated by relying on principles of infrastructure funding and financing that have a sound basis in policy and economics. Many other infrastructure sectors went through decades of evolution and reform to discover the preferred approach. Through careful application of lessons learned in other sectors, the hyperloop can become a reality.

CHAPTER 7. Safety

Primary Author: Ian Sutton

The aviation industry's great efforts into learning from past accidents to prevent future ones has resulted in its current reputation of being very safe. Yet within a four-month period from October 2018 to March 2019, two Boeing 737 Max aircraft were tragically lost. Such events remind us that safety can never, ever be taken for granted.

Each person, company, and regulator in the hyperloop industry should maintain the ethic of making safety a paramount priority. The industry's goal should always be to make hyperloop the safest form of travel possible. This chapter describes a framework to think about safety, discusses major safety issues having to do with hyperloop, and provides initial guidance for the control of major hazards.

The chapter fundamentally assumes that all accidents truly are accidents – events that no one wants to happen. Thus malicious events, ranging from minor vandalism to terrorist acts, are outside the scope of this chapter.

Engineers and managers generally ask three simple questions regarding new technology:

- Will it work?

- Is it safe?

- Will it be profitable?

Addressing the first question, the previous chapters have described the technology behind hyperloop and have shown that, even with issues to address, hyperloop appears to be workable. The chapter after this one considers the third question, looking at economics and profitability. This leaves the central question of this chapter – So, is hyperloop safe?

An initial safety evaluation of hyperloop shows that the technology has features that make it inherently safer compared to other methods of transportation. These features include:

- No at-grade collisions occur because the pods are in a self-contained tube.
- Since the pods or capsules located just a few inches above the base of the tube, they cannot fall from the sky.
- The tubes provide protection from weather events, such as wind, fog, rain and snow.

Some features of hyperloop appear to be less inherently safe than other forms of transport. For example, if a pod traveling at high speed were to deviate from its course, the consequences would likely be catastrophic.

Risk

Risk can never be zero. Any activity has hazards associated with it. Those hazards have consequences, and the likelihood of occurrence can never be eliminated completely. Therefore, a more nuanced way of asking the question would be: "Is it acceptably safe"? That is, in the case of hyperloop, to ask the following:

- Does hyperloop achieve an acceptable level of safety?
- How does the safety of hyperloop compare with other methods of transportation, such as airplanes, railways, and highways?

Before commencing a safety analysis, the elements and management of risk must be understood. Risk has three components: (1) Hazards; (2) Consequences of those hazards (safety, environmental, economic); and (3) Likelihood or predicted frequency of occurrence of the hazards.

The act of climbing a ladder can serve to illustrate risk. The hazard is falling off the ladder. The consequence of the fall can range from a minor injury all the way to a fatality. The likelihood of a fall will depend on many factors, such as the frequency with which the ladder is used, the quality of the ladder's construction, the training of the person involved, and weather conditions.

In general, there is an inverse relationship between consequence and likelihood. For example, when using a ladder, quite frequently people may slip and slightly hurt themselves. However, fatal falls are infrequent. The figure below shows the relationship between likelihood and consequence for hyperloop. The areas of greatest focus for analyses of safety change over time. In the early stages of a project, identifying and controlling high consequence, but less likely, events are the prime focus. As the technology matures, the focus moves leftward in the figure to those situations where incidents occur more frequently, and thus are more likely, but have a lower impact. Currently in the early stages of its technology's development, hyperloop is near the right side.

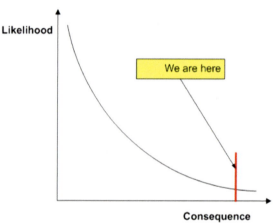

One way of assessing risk is through the use of a risk matrix, as shown in sketch below.

		Consequence			
		Low 1	Moderate 2	Severe 3	Very Severe 4
Frequency	Low, 1	D	D	C	C
	Moderate, 2	D	C	C	B
	High, 3	C	C	B	A
	Very High, 4	C	B	A	A

Events that have low consequence and occur only rarely appear at the top left of the matrix. These events are considered to be acceptable, so they have a risk ranking of "D." The bottom right corner of the matrix shows events that have both a high consequence and a high frequently of occurrence. Such events are never acceptable, and thus have a risk ranking of "A." They are project killers.

With respect to hyperloop in its current state of development, primary interest lies in the 'Very Severe Consequence' column, particularly those that score an "A" or a "B."

Acceptable Risk

Given that risk can never be zero, any safety program must incorporate some means of determining acceptable risk. Although this topic is subjective and emotional, it cannot be avoided. Many factors contribute toward a person's willingness to accept risk. For example, someone may object to having a hyperloop station near her home because it will make the local roads less safe. If, however, she obtains a well-paid job working at that station, her risk tolerance may shift.

At this stage in the development of hyperloop, the most sensible approach may be to first analyze the risk using the technique known as Probabilistic Risk Analysis, and then compare the results with the risk that people take every day as they travel using other methods of transportation. The following statistics describe some of these everyday risks:

- Air travel has one major accident for every 8.7 million flights.
- In the United States, air travel has 0.2 deaths per 10 billion miles.
- For driving, the United States has about 150 deaths per 10 billion miles.

Most passengers will probably expect hyperloop safety to be at least as good as that of the airline industry. Automobile travel does not offer such a good comparison because people feel (rightly or wrongly) more personally in control, and hence are willing to accept a higher level of risk than they would using public transport.

Risk Reduction

We saw that risk has three components: a hazard, its consequences, and its likelihood. Addressing these three items in order generally provides the best way to reduce risk. Returning to the ladder example, avoiding the use of a ladder at all, where appropriate, may be the best way of reducing risk (properly constructed scaffolding is inherently much safer). The next best approach is to reduce the consequences of the event might be able to be reduced significantly. For instance, if a rule requires that a person cannot climb above a certain height, then the worst-case consequences will be minimized. The third, and usually the least satisfactory, means of reducing risk is to reduce the likelihood of the event, in this case, with better training on ladder use or by restricting the use of ladders in bad weather.

Safeguards can be included in risk evaluations. These are items or actions that reduce the consequence or likelihood of an event, but do not stop it from happening altogether. With regard to ladder use, a safeguard known as "tying off," i.e., wearing a safety harness that is firmly attached to the structure while on the ladder, reduces the consequences of a fall.

At this early stage in the development of hyperloop, it is best to focus on eliminating hazards altogether. The next best action is to minimize consequences. Safeguards, such as emergency response, can be considered later during the design process.

While chapter 2 described the commercial appeal of using hyperloop for carrying freight rather than passengers, hyperloop freight transportation, which effectively removes the potential for harm if people are not present when a major incident occurs no one will be injured or killed. By removing people from a hazardous situation, when the worst happens, no one will be injured or killed. Although the economic damage may be substantial, the system remains safe for people. The system is said to exhibit "inherent safety." The idea is that, regardless of the magnitude of the event, the system is always safe, as conveyed in the slogan: "If a man's not there, he can't be killed."

Many of the issues discussed in this chapter, such as loss of directional stability, are critically important no matter who or what is being carried, this chapter primarily discusses the safety of hyperloop systems that transport people. Nonetheless, the inherent safety advantages associated with freight should be kept in mind by way of comparison.

Safety case/safety study. Experience with other industries has shown that a formal Safety Study – often referred to as a Safety Case – serves as an effective way of first understanding, and then controlling, risk. The owner-operator makes the Safety Study "case" to other stakeholders to demonstrate that the system achieves an acceptable level of risk. Those stakeholders include members of the traveling public, people living nearby, employees, investors, and regulators.

A formal safety study provides the several benefits. The study:

- helps define worst-case scenarios and provides guidance on how to reduce the risk associated with these scenarios to an acceptable level.
- provides a basis for comparing the safety of different types of technology.
- provides a framework for developing generic safety programs, thereby minimizing the cost and time required by project-specific programs.
- provides a basis for regulators to develop practical rules, and for engineering societies to prepare standards.
- can help identify opportunities for reducing costs.

Project management. A large project will typically follow a path using the Phase/Gate concept, as shown in the figure below. A safety study can be prepared for different stages, particularly "Concept," "Front End Engineering Design (FEED)," and "Operations."

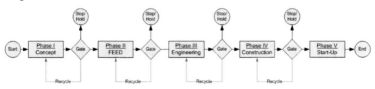

Using the Likelihood/Consequence curve and Risk Matrices described above, the safety study will analyze the system and its associated risks using Probabilistic Risk Analysis and a Major Hazards Analysis, as described next.

Regulations and Standards

In order to ensure a high level of safety, regulators and professional societies throughout the world have issued hundreds of standards and guidance documents. Examples of these organizations are:

- The U.K.'s HSE (Health & Safety Executive).
- The United States OSHA (Occupational, Safety and Health Administration).
- The IEEE (Institute of Electrical and Electronics Engineers) Standards Association.
- The ASME (American Society of Mechanical Engineers).

As hyperloop technology develops, safety will likely be managed mostly through the use of existing, well-established, industry-wide standards. However, at the same time, new standards for major hazards specific to hyperloop technology probably will have to be developed.

The European Union and the USA have started to work on regulations. The EU is creating a Technical Committee in charge of choosing which existing standards would apply to hyperloop, and what new standards should be created. The USA has created a working group with the purpose of regulating new forms of transportation.

The figure below shows the type of regulations and standards that will be developed. They are shown in descending order of priority: regulations carry more authority than codes, which in turn have more authority than industry consensus standards and professional guidance.

Regulations sit at the top level. These legal requirements must be followed. However, regulations often fail to provide needed detail. The next level down, Codes, represents codes and engineering standards written by professional societies. When these are incorporated into law, as often happens, they in effect become part of Regulations.

70

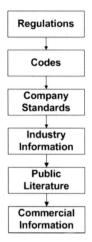

The next level, Standards, includes the standards that large companies will often develop. These provide additional requirements and guidance to supplement the more general codes and industry standards. Next comes Industry Information, representing information shared by companies to ensure that the industry overall is operating safely.

The next two levels are Public Literature – articles, conferences, and Commercial Information – information from private industry that may have a commercial bias, but is often very useful.

Major Hazards Analysis

At this stage in its development, a hyperloop safety program would start with a Major Hazards Analysis (MHA). This analysis identifies hazards that create high consequence events. It could also help evaluate different technologies to determine which approaches are the safest.

This section discusses some of the issues that a Major Hazards Analysis would address.

For each major hazard, it is necessary to determine:

- How might the event happen, and by what causes?
- What outcomes and consequences (safety, environmental, economic) might arise?

- How likely is it that the event will occur?
- What safeguards can be introduced to reduce the level of risk to an acceptable level?

Although this list illustrates the scope of such an analysis, it is not intended to be comprehensive. Examples of some ways to mitigate each hazard are discussed below.

Loss of directional stability. If the pod were to lose control, i.e., to come off its guideway (say, due to a failure in the mag-lev system), it would immediately hit the inner wall of the tube. If the pod were traveling at full speed, the consequences of such an event would be devastating. This situation contrasts with an airplane flight given that, if an airplane loses directional control, it has room to move around in the sky to resolve the problem. With hyperloop, however, there is literally no room for error.

Electrical power failure. Total loss of electrical power would lead to the system shutting down. The pods would quickly come to a halt. Important safeguards include:

- Landing wheels that would deploy automatically.
- Brakes that deploy automatically and would prevent the pods from "coasting" into one another.
- On-board power to keep the passengers safe and to keep instruments operating.
- Stopping all other pods, even if they still have power.

Breach of the tube. Successful operation of hyperloop relies on maintaining extremely low pressures in the tubes. Three types of breach can be considered:

- A pinhole leak, for example through a failed flange gasket.
- A large leak, equivalent to a hole of 1 inch in diameter.
- A total breach — sometimes referred to as a "guillotine break."

Of these, the total breach could have catastrophic consequences. Air would rush into the tube at sonic speeds. A pod traveling along such a breached tube could hit the rushing air, which would be almost like hitting a solid object. Moreover, the high-pressure wave from a breach would travel down the tube in both directions, so other pods could suffer a similar crash-like experience, and secondary tube breaches could result.

To address this possibility, vacuum breakers could be available all along the tube. When pressure sensors pick up a leak, the vacuum breakers would open and fill all sections of the tube with air at a controlled rate. Meanwhile, all the pods would be brought to a safe stop.

Pod depressurization. Two main consequences would result were the pod traveling inside the tube to have a large pressure leak between itself and the tube. First, the pressure inside the pod would fall to near-zero. Therefore, passengers would have to be provided with oxygen masks, as they are now on airplanes. Yet this solution would only be effective for a short amount of time, less than a minute, as the human body cannot sustain a near-vacuum environment. For this reason, the system will require either emergency valves that will quickly re-pressurize the tube, or else the pods will require their own sealing system.

Seismic activity. Seismic activity could cause the tubes to deform or even rupture. (Strictly speaking, seismic events are not hazardous, they are the causes of hazards.)

Events that require emergency response. Should an emergency occur, an emergency response team must be able to reach the event site quickly. Usually straightforward with road and rail travel, access also is not difficult to an airplane once it has landed. But, since hyperloop is a self-contained system, emergency access will be a challenge. Just as regulations on tunnels impose an emergency exit every half-mile, emergency exit portals will exist at regular intervals along the hyperloop tube.

CHAPTER 8. Hyperloop and Climate Change

Primary Author: Stephen A. Cohn

Introduction

Since the industrial era, humankind has created ever increasing global wealth, supported larger and larger populations, and achieved an ever-higher standard of living. This has happened by harnessing the vast stores of energy in fossil fuels like petroleum, coal, and natural gas.

CO_2 and other gasses released into the atmosphere by burning fossil fuels are changing our climate. With rising CO_2, the atmosphere and oceans are warming and changing regional climates. Direct results of climate change include rising sea levels, more frequent and damaging hurricanes, heat waves, droughts, wildland fires, along with declining productivity of farms and ocean biomes, and even extinctions of some species.

About twenty-five percent of human-created CO_2 emission each year arises from transportation. How can hyperloop help redirect our transportation future away from CO_2-generating fuels? For some modes of transportation, harmful emissions can be reduced to insignificant levels by making transport more energy efficient and by restructuring power generation to be largely from renewable energy sources. The most direct path to renewables appears to be using electricity to power transportation, and moving electric generation to methods such as wind, hydro, and solar. Other changes, such as the use of hydrogen fuel cells and biofuels or synthetic fuels (synfuel) can also play a role.

Hyperloop transportation fits within the following three critical environmental strategies:

- *Improve efficiency to reduce overall energy needs.* Hyperloop is very energy efficient. People and freight can be moved using a fraction of the energy of other modes of transport.

- *Move away from gas and diesel, and instead use electricity to power much of our needs.* Parts of the transport sector are slowly transitioning to electricity, with more and more hybrid- and fully-electric cars and delivery vehicles appearing each year. Hyperloop runs on electricity and can be a leader in this transition. This strategy works in tandem with the next strategy.

- *Generate electricity by renewable means rather than by burning fossil fuels.* Energy generated using renewable resources does not add more carbon to the atmosphere. These resources are solar, wind, hydropower, biofuels and synfuels, geothermal, and perhaps nuclear. Run by electricity, hyperloop is primed to lead a positive change to carbon-free transportation.

To emphasize this point, the following figure shows the amount of CO_2 released from using various fuel sources. The specific numbers relate to fuel for rail, but the same message applies to all modes of transportation: Motive electric power from wind, solar, or hydrothermal sources reduces carbon emissions by 98% compared to power from petroleum or coal. Powering transportation using renewables can enormously reduce the carbon footprint of the transport sector.

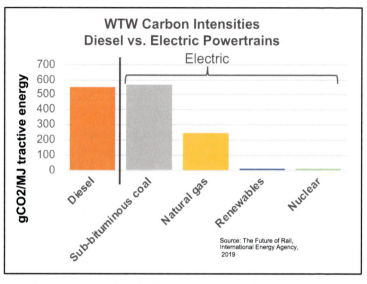

CO₂ emissions (well-to-wheel) to move a train with a diesel locomotive, compared with an electric locomotive using electricity from various sources. "g CO₂/MJ" is grams of carbon dioxide equivalent per megajoule, and tractive energy is the final energy used to propel the train. The wheel-to-well measure includes the carbon emissions due to extraction, refining, and transport of the fuel, as well as the direct emissions during its use. These results are from a 2019 report by the International Energy Agency.

The following sections discuss hyperloop's estimated potential to reduce CO₂ emissions when used as a substitute for selected existing transportation trips. The analysis assumes the presence of a robust hyperloop network on a scale similar to current highway networks, which is a long-term goal. It illustrates where and how large hyperloop can have an impact.

Sources of Carbon Emission in Transportation

Globally, transportation accounts for about 25% of the CO_2 released in the atmosphere according to the International Energy Agency (IEA), which tracks data on all energy sources and technologies. The worldwide total in 2016 was more than 32 Gigatons of CO_2, a number that is growing each year.

Reducing CO_2 emissions enough to have a real impact on climate change will take changes in nearly every energy aspect of our society - from how buildings are heated, to the energy used for industrial processes, to how transportation is powered. The next figure shows the CO_2 contribution from the major sectors of energy use. Transport is among these major sources and represents an area where changes can have great benefit. The figure shows that, within the transport sector, nearly three quarters of global transportation emissions come from road transport (cars and trucks), with the rest of the emissions coming about equally from aviation and ships. Rail transport accounts for just a few percent of emissions.

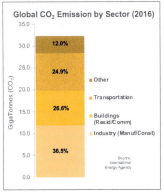

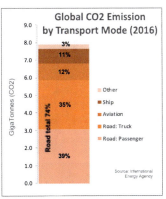

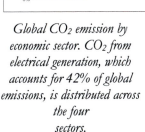

Global CO_2 emission by economic sector. CO_2 from electrical generation, which accounts for 42% of global emissions, is distributed across the four sectors.

Global CO_2 emission by transport mode

Data are from a 2018 report by the IEA.

Avoidance of CO_2 Emission with Hyperloop

If a robust hyperloop network existed, and if it were operated using electricity from renewable sources, what transport emissions might be avoided? This is not an easy question to answer. The answer depends on many data sources, many assumptions, and some educated guess work. But even with uncertainty in estimates, hyperloop's potential to greatly reduce transport emissions remains clear. Below several modes of transport are considered individually to show how hyperloop can make a difference.

Freight trucks. Trucks of various sizes that transport freight, called Road Freight Vehicles (RFV) by the IEA, are central to commerce. Heavy freight trucks, the big rigs of our highways with gross vehicle weight greater than fifteen metric tons, are used for long-haul trips and consume more than half of the petroleum of all RFV.

The RFV category also includes medium freight trucks – typically used for regional trips, but also for long-hauls in some parts of the world – and light commercial vehicles, which may be used, for example, for last-mile deliveries and routine movement of small loads.

A hyperloop network could substitute for some fraction of heavy-freight truck trips, a smaller fraction of medium-freight truck trips, and probably none of the light commercial vehicle trips transporting freight. Studies on diversion of freight transport to hyperloop must consider the cost of moving goods by hyperloop compared to truck. While hyperloop might transport goods more safely and quickly, this likely comes at a higher cost.

How much CO_2 is road freight producing? According to the IEA, road freight produces more than 35% of the transport sector's global CO_2 emissions. Since heavy road freight is responsible for half of this amount, hyperloop could have a very large impact. If the freight from all the trucks we see crowding highways, freeways, motorways, and expressways were instead carried by a robust hyperloop network powered with carbon-free electricity, it could reduce the transport sector's global CO_2 emission by up to eighteen percent. That would be a reduction of about 1.4 Gigatons of CO_2 each year – a huge reduction. Some research already shows potential CO_2 reduction by shifting truck freight to conventional rail or barge, but also recognizes the disadvantage of longer travel times. A shift to hyperloop would not have that disadvantage. Actually, the potential surpasses this because these emissions continue to grow rapidly – they increased by more than eighty percent from 2000 to 2015.

Passenger cars. It was driving – or rather, being at a standstill in traffic – that inspired the 2013 Hyperloop Alpha paper. How much difference would hyperloop make in reducing CO_2 emissions from personal automobile travel (cars and light trucks)? This depends on how many automobile trip-miles could be avoided in cases where hyperloop offers a better alternative. Hyperloop makes sense for longer trips, for example, vacation road trips rather than short trips for commuting or grocery shopping.

To estimate possible impacts, consider trips in the U.S. (where key data is available). According to the U.S. Environmental Protection Agency, passenger cars and light trucks account for about 58% of the U.S. transportation sector's CO_2 emissions – that's a lot! Based on analysis of data from the National Household Travel Survey, almost 92% of U.S. passenger miles are driven for trips shorter than 100 miles. Hyperloop would surely not substitute for these. Longer trips, greater than 300 miles (500 km), represent about 5% of trip-miles driven. This is a modest fraction, but of a very large number. Overall people in the U.S. drive a lot – about 3.9 trillion miles annually.

If all car trips greater than 300 miles were instead made via carbon-free hyperloop, nearly 200 billion miles of U.S. driving would be eliminated. In the U.S., whose average fuel consumption of passenger vehicles and light trucks is about 22 mpg (about 9 km per liter), this means almost 10 billion gallons of gas could potentially be saved each year! This translates to a savings of about 2.9% of all CO_2 from the U.S. transportation sector.

It takes some guesswork to extrapolate this result to the rest of the world. Let's conservatively reduce the estimate from 2.9% of U.S. transport sector emissions to just 2.0% of global transport emissions, since we know U.S. driving falls on the high side. While 2.0% is less than the potential reduction in the trucking sector due to hyperloop, it nonetheless represents a large and significant reduction.

Every reduction of the carbon released into our atmosphere is important. Electrifying long distance road trips with hyperloop also contributes to the overall transition toward electric vehicles. EVs are great for short trips, but drivers suffer from range anxiety and long charging times for longer trips, which can be avoided with hyperloop.

Aviation. Flying is bad for the planet. In general, getting to airports is increasingly inconvenient, seating is often cramped and uncomfortable, and flights have become more expensive as fees are added for "extras" like luggage.

According to the International Civil Aviation Organization (ICAO), aviation accounts for about 2% of global human-generated carbon emissions. Climate researchers and those studying carbon emissions have long recognized that aviation contributes pretty significantly in this way. As one news article notes, "Take one round-trip flight between New York and California, and you've generated about 20 percent of the greenhouse gases that your car emits over an entire year."

ICAO and the aviation industry recognize the need to reduce aviation's environmental footprint. While good progress is being made in electrifying automobile travel, moving large commercial aircraft away from fossil fuels has proven much harder. How much of an impact would moving some commercial aircraft trips away from fossil fuels have? U.S. data from a 2018 NASA report can give a rough estimate.

The left frame of the figure below shows approximately how many U.S. flights are made each day, categorized by trip length. The right frame shows the distribution of flight-miles for each trip length. Which flights might be replaced with hyperloop travel? Given that hyperloop would travel at speeds comparable to, or greater than, jet aircraft, hyperloop would compete with most flights that do not go overseas. Making an approximate assumption that, globally, flights of less than 1100 miles could be replaced with hyperloop, this suggests a reduction of about 60% of flight miles.

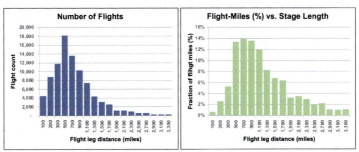

U.S. distribution of flights (left) and total flight miles (right) as a function of leg (stage) length. Based on data from Marien, et al. 2018 (see Further Reading).

Since aviation accounts for about 12% of global transport sector CO_2 emissions, shifting 60% of flight-miles to hyperloop would remove 7.2% of all emissions from the transportation sector – another very large reduction.

In fact, although this analysis uses CO_2 as a proxy for harmful emissions, other aircraft emissions also have a long-term impact on our planet's climate. Jet engine exhaust affects ozone, methane, and water vapor high in the atmosphere, all of which affect the heating and cooling of our planet. The net effect of aviation emissions is complicated, but climate models account for it with a "multiplier" that applies to the direct amount of CO_2. This factor is greater than 2 (and sometimes as high as 10), meaning that the benefit of reducing flight-miles is at least double the CO_2 benefit alone.

Rail. How much can hyperloop substitute for rail transport and its emissions? Actually, quite a lot can be replaced. All long-distance freight and long-distance passenger travel could potentially be replaced by hyperloop, though not local travel. Importantly, most rail emissions come from long-distance transport.

Conventional rail use differs largely by region. Freight transport dominates in North America, while passenger use of rail prevails in Europe and Japan. There are also large differences in energy sources used by trains around the world. For example, freight trains in North America overwhelmingly run on diesel, while both passenger trains in Japan and new high-speed rail in China are electrified.

Analysis of the energy intensity for rail in various regions of the world shows which mode has the greatest potential for CO_2 reduction. For example, per the prevalent uses noted above, in

North America and Europe the energy per passenger-mile needs the most reduction, while in Japan the greatest potential for improvement lies with freight.

Rail stands out as one of the more efficient forms of powered transportation. Worldwide, rail accounts for about 8% of human transportation-miles (around 2.5 trillion rail passenger miles) and 7% of freight transport, yet rail as a whole consumes only about 2% of the energy of the transportation sector as a whole. These numbers include urban metro and light rail, as well as conventional and high-speed rail. A built-out hyperloop network would not replace metro and light rail, but it could substitute for much of conventional and high-speed rail travel.

Data is not available to separate the long-distance and local categories of emissions, but as a rough estimate, let's assume hyperloop can substitute for about 50% of passenger-miles and for about 75% of freight-miles. In addition to shortening travel times, overall this could reduce global rail emissions by 66%. So, given rail's 2% share of transport sector emissions, this suggests an overall benefit of about 1.3%, which represents another significant contribution to emissions reduction.

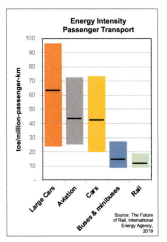
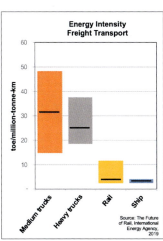

Energy intensity of different transport modes in Tonne Oil Equivalent (toe). (left) Passenger-transport energy per passenger-km. (right) Freight transport energy per tonne-km. Data are from a 2019 report by the IEA.

Ships. Much of the world's products are transported across oceans on huge container ships that are slow, but very energy efficient.

Freight movement by ship accounts for 74% of all freight tonne-kms, yet only 11% of the transport sector's CO_2 emissions. As with transoceanic aviation, hyperloop will not likely substitute for it. Freight movement on river barges could be replaced with hyperloop. However, the cost of carrying freight on river barges is most likely much lower than that with hyperloop pods. Although not zero, we will conservatively assume essentially no CO_2 savings from hyperloop in this transport mode.

This is not to say hyperloop and shipping are independent. As discussed in Ch. 9, they are increasingly seen as complementary, and hyperloop can have a large role in reducing traffic and congestion around ports. When goods arrive at a port aboard container ships, hyperloop would be a natural mode to quickly transport them throughout a region or country. In fact, one of the world's largest port operators is now investing in a leading hyperloop company.

Summary: Transport CO_2 Reduction with Hyperloop, 28%

Putting the estimates together reveals the overall potential of hyperloop to reduce global carbon emission.

The figure below summarizes the reduction in CO_2 estimated for each transport mode. Shifting trips to hyperloop could greatly reduce emissions from freight truck transportation and from aviation. Hyperloop would have an appreciable impact on emissions from car and light truck trips, which are the greatest transportation emitters. Hyperloop would also have a transformative impact on long-distance rail, although rail is already a very efficient mode.

Overall, the total reduction of emissions within the transport sector is about 28%, which would present dramatic, game-changing progress in the global transition away from fossil fuel use. Thus, hyperloop offers a huge opportunity resulting from a single change in transportation technology.

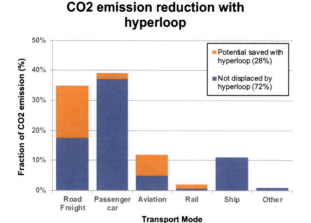

CO2 emission reduction with hyperloop

Potential reduction in CO_2 emissions with hyperloop, by transport mode. The full height of bars is the current CO_2 emission, and the orange area indicates potential reduction by shifting some transport to hyperloop.

Beyond the carbon calculation, there are other insights from this analysis.

First, hyperloop can change the way people travel and the way freight is transported. It adds to, complements, and in some cases substitutes for, existing modes of transport. The analysis suggests that hyperloop could: absorb a large fraction of highway-based trucking, increasing speed and safety; substitute for most domestic commercial flights (but not transoceanic flights), increasing comfort and convenience to city centers; and replace much of the travel on conventional and high-speed regional rail (but not urban rail), again shortening travel times.

Hyperloop can substitute for automobile and light truck trips on highways, reducing road congestion while increasing speed. It would have little effect on oceanic ship travel, although it could greatly change the use of inland barges and port operations.

In summary, although it will take time and resources to create a hyperloop network that is comparable to current highways, doing so would have large environmental, social, and economic benefits. The resulting interconnected world would truly make hyperloop the physical internet.

CHAPTER 9. How will Hyperloop Impact the Environment?

Primary Author: Mary Ann Ottinger, with contributions from Dario Bueno Baques

Introduction

Well planned and implemented hyperloop systems promise great societal benefits. This chapter describes key considerations for hyperloop to be built in harmony with both the natural and human built environment. People have always altered their surroundings to fulfill needs. Globally, human activity and infrastructure has grown over time to a level where the impact of all development, including transport, must be carefully considered. While this was less true when other modes of transportation were in their infancy, it applies very much today to hyperloop development.

Without due attention, developing a hyperloop network has the potential to disrupt the natural environment. This chapter describes issues such as the fractionalization and disruption that accompanies most current ground transportation infrastructure, including noise and vibration that can disturb people and animals, and pollution or chemical contamination, both during construction and under operation. Fortunately, early-stage planning and well-informed decisions can prevent or minimize these potential disruptions for hyperloop.

Hyperloop will impact the built (human) environment and change the way people live, work, and play. Overlaying a new form of transportation like hyperloop in a way that creates synergies with existing forms of transportation will be challenging. The challenge goes well beyond fitting new infrastructure into complex spaces. The economic opportunities and lifestyle changes that are enabled by hyperloop must be planned. Cognizant of the impacts of past transportation advances (interstate highway, high-speed rail, and aviation), this chapter also describes possible economic opportunities from hyperloop, and changes to the nature of urban and rural regions.

Integration with the Physical Environment

Creating infrastructure inherently results in changes to the natural environment, but good planning should strive to preserve as much of the natural environment as possible. Highways, road networks, and train tracks have historically been associated with a problem known as fractionization, where land, habitats, and even entire human communities are separated into "here" and "the other side of the tracks" by roads, railroad tracks, and other manmade barriers.

The Elyria-Swansea neighborhood in the northeast region of Denver, Colorado provides a striking example of this phenomenon. In 1964, Interstate-70 was constructed, bisecting this neighborhood. Pedestrians and cyclists – including students walking to school - can only cross from one side of the interstate to the other at a few locations. To compound the issue, train tracks, along with large industrial sites and the South Platte River, serve to isolate the neighborhood from central Denver. For decades the division has caused the neighborhood to suffer from noise, pollution, and, most pointedly from the effects of a physically-separated community. Fortunately, a project currently underway (more than 50 years after the division) will move this section of the Interstate below ground, reconnecting the neighborhood and allowing for new green space, parks, and bike paths. Hyperloop must strive to avoid fractionalization and improve the connectivity of society without causing harm along the way. Professionals in urban planning, rural planning, landscape architecture, and other aspects of human geography can apply their expertise to integrating transportation infrastructure with the natural and human environment.

Hyperloop's ability to minimize fractionization depends on where it is elevated, whether on the surface or if it is located in tunnels that don't extensively disrupt the surface environment.

Elevated hyperloop infrastructure. Hyperloop infrastructure has a weight advantage in that it carries single pods that are much lighter than a roadbed or train tracks. Lighter weight provides the option of elevating tubes above ground on pylons or other pillars. The need for a long, straight alignment also favors an elevated structure, where varying the height of the pillars keeps the tube level despite irregularities at the ground level. Also, an elevated structure does not impact existing surface infrastructure (roads, train tracks), disrupt agricultural activities, or disturb pedestrians, cyclists, and wildlife that can still pass beneath the tubes. Much of the hyperloop architecture envisioned by commercial companies includes this elevated concept, shown in the figure below.

Hyperloop located on an elevated path traveling over existing surface infrastructure.

Surface-level hyperloop configuration for above-ground transport. A hyperloop line at surface level, on the other hand, will present the same unfortunate challenges as surface level highways and railroads. However, if there are opportunities to run the hyperloop tube alongside existing routes, this at least would not cause additional division. Some rural and wildland surface routes have mitigated surface fractionization by providing crossings for people and wildlife. For example, wildlife crossings are included along the Trans-Canada Highway (see figure below) that allow animal populations to remain connected. Studies have shown that animal populations fragmented by human barriers have smaller gene pools and lower survival rates. Planning for wildlife corridors and human crossings as part of the design of any surface-based hyperloop can reduce such harm from fractionization.

Wildlife corridor (Ecoduct) crossing over a highway.

Underground hyperloop. Hyperloop designers can choose the infrastructure configuration best suited to each part of a route, including the choice of an underground. configuration. Hyperloop lines in underground tunnels are, of course, the least intrusive on the landscape.

Similar to rail, underground segments may be used to pass through mountainous terrain, while keeping the track relatively level with access near cities. Surface level infrastructure with crossing opportunities may be reasonable for sparsely-inhabited regions – consider desert areas. However, for many long segments, infrastructure elevated on pillars probably offers the best compromise of utility and expense. Impact studies will determine more precisely the potential effects of hyperloop lines on their environment. Hyperloop designers and promoters must plan for integrating infrastructure with the natural and human environment better than what has been achieved to date.

Noise and Vibration

Noise from trains, highway traffic, and airplanes can disrupt people's ability to work, play, and enjoy life. People who live near jet airplane approach and departure routes complain of noise, and sometimes municipalities restrict airport hours because of it. As developers build more housing developments very close to highways, noise mitigation strategies (large walls near the highway, thicker windows in homes) are proving only partly effective. Such issues often present a dilemma. Although many communities have raised exceptions to trains blowing their whistle at crossings, finding the noise too disturbing, the elimination of whistles would make crossings more dangerous.

Will hyperloop create similar disruption? Environmental concerns have been raised that noise or vibration produced by the movement of hyperloop pods within the tube may disturb its surroundings. Experience at existing test tracks in the U.S., along with simulations, show so far that hyperloop will not create much noise pollution. Designers explain that, since the system is levitating, no noise emanates from the contact of wheels on a track. Moreover, because the system operates in a vacuum, the noise that would be created by the friction between the pod and air is mostly avoided, in turn reducing the propagation of sound (to quote the technically-correct tagline from the film Alien, "In space, nobody can hear you scream"). Vibrations also may be reduced because the vehicles will be much lighter than a train, and because magnetic levitation can absorb vibrations much better than traditional suspension systems. These explanations do not remove the need for thorough studies and full-scale tests to measure precisely the noise and vibration hyperloop lines may generate and if needed, engineer ways to reduce the potential for environmental impacts.

Pollution and Chemical Contamination

All forms of transportation pollute the surrounding land and atmosphere to varying degrees. Tailpipe emissions from cars and trucks add carbon to the atmosphere. Aircraft pollute the upper atmosphere where the environmental harm may be multiplied. In winter the salts and other substances used to treat icy roads contaminate the surrounding soil and streams. Many harmful chemicals are found along rail lines, including creosote used to treat railroad ties and the lead and arsenic from coal ash and cinders. Heavy metals and dangerous chemicals from leaks are also found along lines.

Because hyperloop construction and operations will require use of some chemicals as well, such as anticorrosion coatings on the surface of tubes, it's important to safeguard the surrounding environment from these potential chemical contaminants. This can built into the system concept and design, which should emphasize less harmful choices for needed chemicals and establish a robust monitoring program. Such a program should be fully integrated with the procedures that monitor and ensure safety, operations, and maintenance programs.

While above ground (surface and elevated) architectures expose the local environment to greater risk, contamination underground with tunnel-based hyperloop infrastructure should not be ignored. Underground hazardous chemicals can harm local animal and plant species, transport to water sources, and present a hazard for any future use of that land. As recent experience with industrial spills and redevelopment of some land has shown, it is difficult and extremely expensive to clean up contaminated land.

The potential environmental impacts of chemicals should be considered in terms of their lifecycle as they flow through ecosystems, potentially with effects at different points along the way. Risk and potential adverse impacts should be managed according to the characteristics of the chemicals involved.

The past use of polychlorinated biphenyls (PCBs) in electrical infrastructure offers a cautionary tale for hyperloop. PCBs are very stable compounds that persist in the environment (in soil and water) for many decades following a contamination event. They are also endocrine-disrupting chemicals, which mimic our own hormones and as such have the potential to be extremely harmful to humans and wildlife. PCBs originally entered the ecosystem from industrial chemical production and from the production of transformers. Because PCBs were often used as lubricants due to their ability to withstand high temperatures, they are extremely persistent in the environment. Such potentially hazardous chemicals can be avoided for hyperloop, and the selection of all components should maximize environmental safety. For example, the hyperloop infrastructure is mainly composed of a steel tube, with chemicals applied on the exterior of the tube to prevent corrosion, deterioration, and ultimately, to prevent air leaks. These chemicals should be strategically selected to ensure minimal loss over the long-term into the environment. Care should be taken with welding, vacuum pumping stations, and emergency exits, which can contaminate the surroundings and can generate pollution during construction. Designers should understand all pollution – from big and small sources – and ensure there are no harmful impacts.

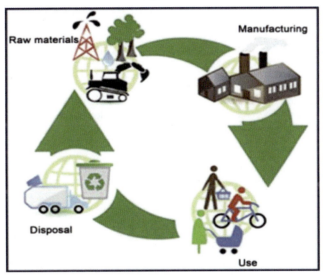

Representation of the lifecycle of compounds. Once manufactured, they are used and then disposed of, allowing reentry into the environment. From the "Risk Management Sustainability Technology," U.S Environmental Protection Agency.

Environmental Disruption During Construction

As with any infrastructure construction, environmental conservation goals apply to the construction phase of hyperloop. For example, during construction the area in which the tube is being built is unavoidably disturbed. Wherever possible, reducing these impacts should be a consideration in construction. For example, if infrastructure is on raised pillars, the construction might be optimized by first manufacturing prefabricated pillars and tube sections in factories, and then assembling them on site. This can reduce construction time and limit the environmental impact of construction. A temporary road would still be required for the excavators, trucks, and cranes needed to dig the foundations of pillars and erect the superstructures, but the road could be decommissioned after construction and the land could be returned to its original function.

Environmentally sound construction practices include consideration of the source and transport of materials. The entire life cycle of environmental impacts should be considered. As shown in the figure below, raw materials needed for construction are first refined into products and tools by manufacturing. How much energy is needed to source all materials? Does producing the materials harm the environment? Are there better construction methods to reduce the environmental footprint of maintenance? How can design choices enable recycling or safe disposal of materials at the end of their useful life?

Local and Regional Economic Opportunities

As shown in the figure below, the hyperloop station concept, like that of many transportation hubs, offers easy access to passengers, transport efficiency, and pleasant surroundings during transit, along with amenities associated with the station. These on-site characteristics help to ensure public acceptance and comfort, adding to the clear on-site economic potential of transportation hubs.

Economic development around transit stations is well established. New stations within cities attract nearby commercial development or redevelopment. For example, an economic analysis of Denver's investment in a regional rail network found huge transit-oriented development around new stations - nearly 18,000 residential dwelling units, 5.3 million square feet of retail space, an equal amount of office space, and 6.2 million square feet of medical space were built within one-half mile of transit stations from 1997 to 2010. Development and renewal benefit the local economy and also generate government revenue through greater sales tax and real estate tax. At the same time, impacts such as congestion, gentrification, and possible loss of a sense of community should be considered and mitigated.

Hyperloop station concept visualization. (credit: Student work from the 2014-15 IDEAS Mobility Studio: Hyperloop with Craig Hodgetts, Marta Nowak and David Ross. Image courtesy of UCLA Architecture and Urban Design).

At a larger scale, hyperloop can lead to a cluster effect benefiting both local and regional economies. Clusters are geographic concentrations of businesses that are interconnected and gain from proximity to each other. Access to efficient transportation is a key to the development and location of clusters. These concentrations encourage innovation, exchange of ideas, competition, and ultimately business development. Because of hyperloop's speed and efficiency, hyperloop transport between two business centers may encourage the creation of a larger cluster and provide a substantial economic boost to the connected regions. Hyperloop can connect major existing clusters and promote regional integration.

Impacts on the Shape of Communities and Industries

New technologies have always shaped and reshaped communities. As described in Chapter 2, development of the highway system and the airline industry both dramatically changed long-distance mobility. They changed where industries could locate and thrive, expanding some markets and consolidating others. They even impacted family dynamics, as many in a new generation were able to pursue careers farther from their roots, but still remain connected with family that was many miles, yet still only a day's journey, away. And, of course, aviation brought continents closer together – an hours-long flight replacing a maritime journey of days or weeks. These changes created great advantages for those cities and towns with easy access to the new routes and markets. But they also left behind bypassed places with less access. Many towns along old Route 66 are nostalgic, but mostly disadvantaged economically given they are not on an interstate exit.

Similarly, a hyperloop line will encourage people to live further away from their workplace, perhaps enjoying a more affordable and more comfortable lifestyle, while still keeping the same commute time. According to the Marchetti constant (an idea proposed by Italian physicist Cesare Marchetti), people's willingness to commute is limited by commuting time rather than commuting distance. Thus, with faster transportation, people can live farther from work. When Japan built the Shinkansen high speed train lines, some new stations were in less-populated areas, but growth was spurred around the stations when commuters moved into these areas. As the hyperloop network expands and stations are built in the countryside, growth will need to be carefully planned.

Urban and rural areas are typically impacted differently by transportation development. Urban planners can actively influence how an urban space evolves and consider its connections to suburbia. This includes careful attention to road networks, public transportation, and commercial zoning. Planners are challenged to enhance positive trends, like quick, easy, and affordable access, while countering negative trends, such as overcrowding or regional sprawl.

Rural environments, on the other hand, can be more vulnerable to unintended consequences from growth and changes in mobility. Social implications need to be carefully considered. Growth is generally considered good, but not when it forces unwanted change, strains limited resources like available water, removes unique or needed habitats for wildlife, or even impacts climate by displacing premier biologically-productive ecosystems such as rainforests. The ability of the region to plan and control growth with due consideration given to the interests of local towns cannot be ignored by planners and hyperloop companies; this new mode of transportation should fit everyone's needs.

A clear strength for hyperloop would be to serve as a catalyst for regional integration – not only connecting major cities and hubs, but also bringing connectivity to smaller communities. Current intercity rapid trains carry hundreds of passengers, but do not stop at every town to avoid adding hugely to the trip time for everyone. Even with just a few stops, all passengers are inconvenienced, and only a fraction of the passengers embark and disembark. Air travel is the most restrictive means of transportation since aircraft need a runway and terminals. Hyperloop promises to bring great flexibility - a high frequency of small vehicles will be able to bypass a station if no passenger needs to stop at it. This encourages having many stations along the line, helping to make regional integration a reality. However, building a station has a cost, and the number and location of the stations along the line has to be carefully planned to keep the project financially viable. This is why spatial planning and regional development policies must be central to any hyperloop development project. Local planners and the local population must decide how best to integrate a hyperloop station into their town, and how best to encourage or limit associated growth.

Promise and Complexity of Underground Hyperloop in Cities

Hyperloop construction within cities quite likely will take place underground as for a subway, and, once built, functions least disruptively for a city. In addition to keeping people out of unpleasant weather, underground development has offered demonstrable benefits. Consider the vast network of subway-connected underground shopping areas in central Montreal, which can be a welcome respite for shoppers from the cold and snowy winter outside.

However, as above ground, construction of infrastructure under cities, brings its own complexity. For example, consider the Crossrail Project in London to build the Elisabeth line, a new subway line that will cross all of London from east to west and allow efficient connections to Heathrow Airport. This project is due to be operational in 2023 (much delayed from original goals). The requirement that several above-ground structures in the center of London be removed has resulted in construction that has inconvenienced many individuals and businesses. The experiences associated with this and other infrastructure projects should be applied to minimize delays and inconveniences for hyperloop construction.

On the positive side, the design of new above-ground stations and structures will provide an opportunity to include modern and unique architectural elements to herald the forward-looking and transformational system within.

Summary

Hyperloop will impact both the natural environment and our cities and towns. As with any large infrastructure project, if not carefully planned, it can cause great harm. However, with the combination of sound design by professional urban and rural planning professionals, the involvement of local communities and all stakeholders, and learning from past transportation projects, sources of harm can be mitigated or avoided while positive impacts can be maximized.

CHAPTER 10. What Hyperloop Means to Security and Resilience

Primary Author: Dane S. Egli

Introduction

These are historic and consequential times. Advances in technology and seismic globalization have led to innovations that inspire our generation with excitement and hope. Hyperloop is one such innovation. But the world also increasingly faces challenges enabled by technology. New ideas are needed to counter terrorist attacks, trans-national crime, cyber breaches, and other security threats. Challenges extend internationally and are aggravated by economic instability, geopolitical pressures, and even structural kleptocracy. Threats and their responses collectively influence and reshape supply chains, trade markets, and the global economy. These threats affect all infrastructure and modes of transport, including hyperloop.

This chapter describes the importance of hyperloop for homeland national security planners, policymakers, and those focused on building a more resilient transportation capability in the 21st century. It describes how the unique characteristics of hyperloop will have a significant role in strengthening any nation's safety, transportation, economic, and environmental objectives. Hyperloop offers a potential source of security where security is needed most—within complex, interdependent, critical-infrastructure transportation systems.

Four connected factors link hyperloop to improved security: (1) hyperloop as an emerging form of technology that adds new capabilities to the existing intermodal transportation network, (2) hyperloop's contribution to critical infrastructure resilience by offering redundancies in the form of speed and safety, (3) the role of transportation infrastructures as critical ligaments of any vibrant economy, and (4) the foundational importance of the economy as an essential precursor to a nation's security. Together, these four interconnected elements provide a justification for building high-speed tube transportation.

A transportation technology that enables critical infrastructure resilience can be directly linked to security and to a vibrant economy. The diagram below illustrates the simple taxonomy.

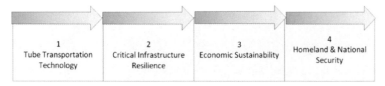

1	2	3	4
Tube Transportation Technology	Critical Infrastructure Resilience	Economic Sustainability	Homeland & National Security

Hyperloop-Resilience-Economic-Security Continuum.

The Transportation – Economy - Security Nexus

The most compelling feature of high-speed tube transportation is its ability to transport goods and passengers to their destinations much more quickly, while relieving congested roads, railways, rivers, ports, and runways. It thus introduces green, safe transport to a complex interconnected supply chain, infusing a new level of resilience and growth to the economy. But it actually offers much more. Adding tube transportation to the existing intermodal network can enhance the resilience of the combined transportation system and also enhance operational security. Hyperloop adds redundancy, flexibility, and sustainability because of its rapid and all-weather delivery of commodities, freight, and goods to market. This would complement existing – and oftentimes fragile – highway, rail, pipeline, aviation, and maritime travel and shipping.

Because homeland security is fundamentally founded and enabled by a strong economy, this new transportation system, developing eventually into a network connecting countries and regions, acts as both a consumer *and* producer of resilience. It supports the arteries of local and regional economies, and thereby strengthens community security and resilience.

A nation needs a vibrant economy to project power in the form of diplomacy, international engagement, coalition-building, financial influence, business collaboration, and free trade. The movement of commodities, today primarily by shipping containers (see Chapter 8), lies at the heart of a strong economy. Ideally this movement relies on an efficient, reliable, speedy, durable, and safe conveyance that connects ports, cities, and communities. Hyperloop offers a valuable addition to the intermodal transportation system because of its ability to complement and reinforce existing transcontinental rail, integrated trucking, aviation cargo, and maritime shipping networks. Therefore, if it can be commercialized and monetized in a way that is practical, effective, efficient, and affordable to the public and private sectors, hyperloop will support and strengthen this transportation-economy-security nexus.

A review of the history of the United States interstate highway system – envisioned, designed, and constructed during the Eisenhower administration – offers interesting points of comparison. The authorization for the Interstate Highway System came through what is popularly known as the National Interstate and Defense Highways Act of 1956. The word "Defense" in this name recognizes the security and resilience aspects of transportation. The original network was completed 35 years later, and has since been extended. In 2016, it had a total length of 48,181 miles, and an estimated one-quarter of all vehicle miles driven in the United States use this system. As was the case for this highway network, the hyperloop transportation system will likely require public enthusiasm and support. It will take government legislation, in all likelihood a combination of commercial investment and financial subsidies, and senior political champions. Supporters must understand that hyperloop is about more than just transportation infrastructure and speedy trains. Hyperloop is a national security and national resilience enabler that will make our economy stronger due to its greater speed, durability, and access to the rest of the country.

The Maritime Ports Example: A Proof-of-Concept

In the U.S. the majority of the national economy depends directly upon movement of freight, goods, and commodities to and from strategic seaports. Hyperloop's integration into the maritime sector deserves serious consideration when studying its economic connection to security and its potential to enhance security and resilience.

Since more than 80-percent of world trade is carried by sea, maritime transport constitutes by far the most important means of transport of goods. This means of transport has been growing annually by around 3.1% for the past three decades. Ensuring this sector remains resilient to hazards and threats requires robust and adaptive solutions such as hyperloop. This section uses the maritime domain as just one example to discuss the rich opportunities, as well as potential barriers to entry, when making the security value proposition to an existing sector.

If containers and bulk cargo can be moved rapidly from ports of entry to a distribution center further inland, this will free up the valuable multi-purpose waterfront real estate currently needed for port facilities and cranes. This movement of shipping containers could be accomplished by connecting offshore transfer stations to an inland intermodal shipping hub (like some major ports in the world, including Los Angeles/Long Beach, Singapore). These inland distribution centers then become the transportation hubs for trucking, rail, and air shipments rather than continuing to funnel heavy transportation traffic into the urban seaport waterfronts. This is not only a potential answer to port congestion, lagging freight mobility, and excess carbon emissions, it creates a dispersed security zone. Goods are moved quickly away from a single chokepoint in the maritime harbors, where existing transportation infrastructures and cargo are highly vulnerable to terrorist attacks, flooding, or storm surge. This is just one way hyperloop could support transportation security.

The U.S. has some 360 ports and 23,000 miles of navigable waterways. A high-speed, highly reliable network that quickly delivers goods and passengers would enhance the current transportation infrastructure—maritime in particular—providing redundancy and sustainability. Hyperloop has a unique ability to improve the speed, efficiency, capacity, and effectiveness of the maritime transportation system and the flow of containers in maritime trade. This will contribute directly to the economic strength of the nation and provide an important economic stimulus in the form of jobs, trade, and market activity.

The U.S. intermodal freight transport system integrates more than $17 trillion of the annual U.S. economy. Of this, maritime transportation accounts for more than $4.6 trillion of commerce each year, supported by 1,200-1,500 commercial ships entering major maritime ports from Seattle, to Los Angeles/Long Beach, Houston, Norfolk, and New York City *each day*. As the figure below shows, trucks carry most of the tonnage and value of freight within the United States, but railroads and waterways carry significant volumes over long distances. Most of that cargo, freight, and goods must enter the system through a maritime port.

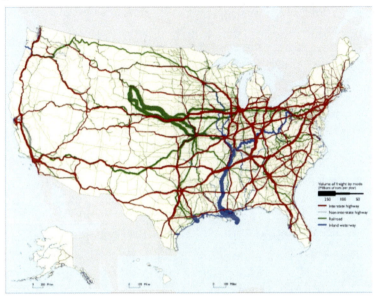

Freight Flows by Interstate Highway (Red), Railroad (Green), and Inland Waterways (Blue).

The potential role of high-speed tube transportation, integrated into and supporting maritime ports, requires deeper analysis and clear justification. The following facts, based on a report by the United Nations Conference on Trade and Development, underscore some important variables that will help planners and decisionmakers understand the value proposition and risk factors associated with hyperloop and other technological advances in maritime ports.

103

- Maritime transportation is the backbone of international trade and serves as a key performance indicator for economic growth due to trade activity and job creation. Global trade and supply chains are heavily dependent on well-functioning port systems.
- Around 80-percent of global trade by volume and over 70-percent of global trade by value are carried by sea and are handled by ports worldwide.
- Vessels carry the largest share (in terms of value) of U.S. export goods annually.
- Advanced technology has become a crucial element in ports, transforming how shipping operations are conducted; New technologies, including autonomous ships, drones, and distributed ledger technologies (e.g. blockchain) promise to further increase efficiency and reduce costs.
- Uncertainty remains in the maritime industry regarding the safety and security of new technologies, including concern about cybersecurity vulnerabilities. In response, governments and the maritime industry are continuing to improve the safety and risk management culture for ports, and are adopting legal, policy, and regulatory frameworks to the needs and risks of technological advances.

The table below, although not all-inclusive, summarizes factors and metrics that must be addressed with any technology change. They are based on items raised by the maritime transportation community, and they are key factors to consider in any plans to integrate hyperloop into national and global transport systems.

Factor/Metric	Key Impacts
Economics	Maritime transportation is the key to global economic growth and job creation.
Measures	Operational, financial, economic, environmental, and social performance.
Trade Markets	Global trade (exports and imports) transported via maritime transportation.

Effectiveness	Advanced technology improves the way shipping operations are conducted.
Efficiency	New technologies offer increased efficiency of operations and reduced costs.
Technology	Example: Distributed ledger for systemic savings and security improvement.
Vulnerabilities	Must address system-wide safety, security, and cyber vulnerabilities.
Risk Management	Risk management culture to comply with legal frameworks.
Integration	New technologies need to be interoperable internally and externally.
Governance	Legal, policy, and regulatory frameworks must be applied at all levels.

Impact Factors for Transportation Sector Changes

These impact factors highlight strategic criteria that private sector investors and government planners would need addressed with data, metrics, and analytics before making decisions to support any new transportation technology, including hyperloop. As technology readiness improves within hyperloop companies through feasibility studies, test tracks, and prototype testing, planners must develop specific case studies and simulations that inform a proof-of-concept demonstration that the system performs as advertised. For this maritime case, there will need to be an operational technology demonstration at a maritime port, measuring the return on investment and ability to significantly improve port congestion, freight mobility, and cargo movement.

Complex Interdependencies Increase Risk

Our society relies upon sixteen critical infrastructure sectors every day to sustain our livelihood, workforce, and the movement of critical goods, services, and commodities. These include major lifeline sectors for water, power, communications, transportation, internet, public safety, and medical services. This system of systems has always been *interconnected*, but in the 21st century—more than ever—they are increasingly *interdependent* upon one another. This means that our marketplace provides more opportunities for both "makers" and "breakers." Favorable conditions exist for *makers*—start-ups, new investors, and market dynamics enhanced by the internet-of-things and open architectures with rapid collaboration –yet those same conditions simultaneously provide a target-rich environment for *breakers* – nefarious actors engaged in cyber terrorism, identity theft, technology corruption, and stealing intellectual property. Further, these network interdependencies introduce the prospect of cascading failures. A small local mishap can trigger a subsequent series of infrastructure dominoes to fall, leading to regional or national-level system disruptions.

In the transportation infrastructure sector, interdependency risk can be reduced by offering a resilient alternative in the form of high-speed tube transportation to counter growing threats, vulnerabilities, and consequences. Hyperloop design offers not only fast, safe, economical, and environmentally responsible transport, but also a relatively more independent and sustainable transport option. Fundamentally, hyperloop technology will make a nation's infrastructure system more resilient because it will help the nation's infrastructure system withstand and recover from disruptive events, reducing the severity of the impact when inevitable crises occur.

The creation of new transportation alternatives such as hyperloop also provides a catalyst for improving safety and resilience by operating in concert with other new technologies. New technologies have emerged and advanced in the past decade at a rate never imagined—robotics, 3D printing (even food printing), autonomous vehicles, super-computing, data analytics, quantum computing, and artificial intelligence to name a few. They are high-tech harbingers providing examples for what can happen in the high-speed tube transportation area. Leveraging the capabilities of hyperloop and other emerging technologies can create a higher level of community resilience, thereby enhancing local, state, regional, and national security.

Hyperloop and National Defense

Consider the strategic military-security requirements of nations that face both nation-state and non-nation-state enemies in the 21st century asymmetric threat environment. In the U.S., the Pentagon's Third Offset Strategy considers how to use technology, science, and innovation to retain the advantages of a world superpower through innovation and modernization. Transportation technology and advanced mobility capabilities are vital national security interests that contribute to any nation's military, economic, and transportation strength. This is where hyperloop technology can potentially be leveraged to inject innovation and to modernize a national defense structure.

Tube transportation has the potential to create strategic advantage well beyond peer competitors. To draw analogies from the past, consider the 1950s, when the unique potential of tactical nuclear weapons made the U.S. an unrivaled power in a world of dangerous threats. This military and technological advantage was instrumental in maintaining peace. A similar asymmetric advantage in the early 1990s came from development of precision-guided conventional weapons that allowed previously unachievable pinpoint accuracy. This technological breakthrough also became a strategic deterrent.

Hyperloop technology can have a similar positive impact in the hands of peaceful nations by offering economic stability and national security advantages in the 21st century. As already discussed in the example of maritime ports, the rapid transport of goods and services strengthens supply chain networks and complex interconnected critical infrastructure sectors that thrive on speedy conveyances and systemic resilience. Consider the potential for rapid and reliable shipment of humanitarian aid, nuclear materials, and transport for organ transplants or acute medical treatment. These types of transportation, logistics, mobility, and deployment capabilities are needed by a modern armed force as well as by society at large.

As we learned during the past century, strength and peace can be achieved through economic, joint military, and technological superiority. Hyperloop's strengthening of intermodal transportation is a strategic advantage because it offers the ability

107

to move resources, people, medical supplies and sensitive material to key locations in minutes versus hours, and hours versus days, thereby supporting local and regional economies in the face of natural or man-made disasters and security threats.

Summary

While many questions remain about the final designs and passenger response to riding in hyperloop, creative and detailed concepts have been proposed. Engineering, biology, architecture, and business considerations will combine to create the most feasible, effective, safe and comfortable pod designs for human passengers.

As described throughout this book, the science, technology, and engineering elements of hyperloop transportation are quickly advancing. But developers will still be challenged by uncertain political will, access to right-of-ways, and economics.

To translate the technology and excitement into real improvements for transportation, including those for security and resilience, requires a consensus among government leaders, municipality officials, political leaders, and private-sector investors, as well as the wide range of international hyperloop companies. These stakeholders and beneficiaries must join forces and integrate designs to form a seamless system. These will need to agree that tube transportation is not just a win for economic markets, opening work-commuting options for the workforce and sustaining regional flow of goods, but also a potential gain in security and resilience that benefits all of society.

CHAPTER 11. Companies and Projects: The Current State
Primary Author: Thierry Boitier

Rather than representing a whole new concept, hyperloop is instead a new word denoting the rejuvenation of an existing concept – one that emerged in Elon Musk's Hyperloop Alpha paper. His personal enthusiasm triggered a surge of activity, motivating engineers to begin and join related design activities. It captured the interest of students, companies, governments, and the general public. This chapter describes the current state of what is called the "modern hyperloop" enterprise – its participants, activities, advanced studies, and forthcoming projects, from the Alpha document (2013) to the first-ever ride for two passengers in a hyperloop system (2020).

Students and Hyperloop

Student engineering competitions on many topics and areas of engineering are organized each year. They encourage students to learn and to use their skills and knowledge developing solutions to real world issues. They also introduce the latest academic concepts and methods to industries. In 2014 SpaceX launched an annual engineering competition for students to develop hyperloop. Students have faced more advanced challenges and objectives each year.

The first competition, in January 2016, focused on conceptual design and prototyping, and attracted 36 teams from four continents. It rewarded several teams for their innovative designs. A second phase of that competition, held one year later, evaluated team prototypes. Twenty-seven teams entered pods that underwent a series of rigorous tests, especially for functionality and safety. The final test entailed a speed run in near vacuum using a one mile long, six foot outer diameter tube installed near SpaceX headquarters in Hawthorne, California. WARR Hyperloop (University of Munich, Germany) won the competition with a top speed of 58 mph, and Delft Hyperloop (Delft University, Netherlands) won the Best Overall Design prize.

The second, third and fourth competitions were held during the summers of 2017, 2018, and 2019. The challenge each year focused on top speed, but also involved passing safety and function tests. Each competition required the students to make new developments. The criteria for the second competition was "maximum speed with successful deceleration (i.e. without crashing.)" The third competition called on teams to develop an onboard communications system (previously provided by SpaceX) and to be fully self-propelled (SpaceX had previously provided an external pusher). For the fourth, most recent competition they also were required to stop within 100 feet of the end of the tube. At this competition the TUM Hyperloop pod (formerly WARR hyperloop), Technical University of Munich, Germany reached 288 mph. The fifth competition was expected to happen in summer 2020, and Elon Musk said that the "competition will be in a 10km vacuum tunnel with a curve." However, no official competition has yet been organized. But another 2020 initiative did emerge - the "Indian Hyperloop Pod Competition," organized by the Indian Institute of Technology in Madras. Unfortunately, because of the Covid-19 pandemic, that competition was postponed.

About 20 teams participate in these events each year. The SpaceX Hyperloop Pod Competition series has helped popularize and energize the key technologies. It attracts a large media presence and coverage, and the student contributions spread technical ideas and knowledge that can more quickly advance the hyperloop industry's development. For example, four out of the six main companies that formed to develop vacuum-train technologies grew out of student teams that took part in Hyperloop Pod Competitions—a testament to the impact of these competitions. Aside from these direct engineering competitions, students of all ages have chosen hyperloop-related projects when their classes have given them the opportunity. These projects may be about hyperloop technology, the societal impact of very high-speed transportation, or even business studies exploring the profitability of the hyperloop.

The Private Sector and Hyperloop

Quite a few companies have announced that they are working on hyperloop core technology, route development, or associated areas such as the passenger onboard experience. Six private sector companies have at least some publicly available information on their fundraising. They are engaged in hardware development with test plans, and are active in working with governments toward a regulatory framework. This chapter summarizes their origin and status.

The first was founded in 2013, shortly after publication of the Hyperloop Alpha paper, and the most recent one formed in 2017. Perhaps the most surprising fact about them is that all are "start-up" companies. Existing, well-established rolling-stock manufacturers are not (so far) visibly exploring tube transportation technologies, but focus their research and development efforts on improving existing technologies.

These six companies face the prime challenge (and opportunity) of starting from a semi-blank page and merging technological innovations with the technologies and engineering advances of earlier vacuum tube transportation efforts. Although all six follow established project development processes, their internal organization and industrial partnerships vary.

Listed by date of incorporation, the six companies are HyperloopTT, Virgin Hyperloop, TransPod, Hardt Global Mobility, Zeleros, and Nevomo. The information reported here is current as of December 2020.

HyperloopTT. HyperloopTT, also known as Hyperloop Transportation Technologies, was formed on November 4th, 2013 in Los Angeles, California (USA), shortly after the publication of the Alpha paper. The company is using the online platform JumpStartFund to coordinate its development of technology. JumpStartFund is a crowd collaborative platform initially created by a co-founder of HyperloopTT. They claim to have more than 600 experts around the world who work at least 10 hours per week in exchange for stock options. HyperloopTT positions itself as an integrator of different technologies, rather than building their own solution.

In 2018 the company opened a research center in Brazil, and in 2019 they completed construction of a 320 m long full-scale test track in Toulouse, France. HyperloopTT reports that a pod has been assembled in Spain and will soon (as of 2019) be tested on the full-scale track in realistic environmental conditions. HyperloopTT has reportedly received funding of USD $30M, from EdgeWater Investments.

Virgin Hyperloop. Virgin Hyperloop, initially called Hyperloop Technologies and later Hyperloop One, was created in 2014 in Los Angeles, California (USA). In addition to their headquarters in Los Angeles, they have built a factory and a test site north of Las Vegas, Nevada. The company conducted its first "Propulsion System Open Air" test in May 2016, accelerating a light pod to 136 mph (219 kph) in 2.2 seconds. In 2017, in a 500m long, 3.25m diameter tube called "DevLoop" the company began technology tests with a full size pod in a partial vacuum. The first test, described as a "Kitty Hawk Moment" was conducted on May 12th, 2017. Over the next few months, the prototype reached a maximum speed of 192 mph (309 km/h). The state of West Virginia has been selected for hosting the next "Certification Center". On November 8th, 2020, Virgin Hyperloop completed the first-ever human ride in a hyperloop system. Josh Giegel, Co-Founder and Chief Technology Officer and Sara Luchian, Director of Passenger Experience travelled at 172km/h (107mph).

Reportedly, Virgin Hyperloop has received USD $448M, in 6 rounds, from a variety of investors. These include venture capital firms, industrial groups such as DP World, Virgin, SNCF, and others.

TransPod Inc. TransPod Inc. was co-founded in Toronto, Canada on October 9th, 2015, with the intent to design a tube transportation system that would be more advanced than the hyperloop concept as initially introduced. TransPod aims for a holistic design based on physical principles that integrates innovation, technology, operations, safety, and profitability.

Through its patents and its research-oriented secondary website, TransPod presents its solution to reducing the cost of infrastructure by removing the need for electromagnets on the track.

The company secured an initial investment from an Italian industrial group in 2015, leveraging opportunities for future partnerships, and has activities in Canada, Italy, and more recently in France. In 2018 TransPod received government approval and a construction permit for a half-scale test track north of Limoges, France. Construction was initiated in 2020, before being halted because of the Covid-19 pandemic. In 2019 it released the first fully publicly-open report on a hyperloop preliminary study. This study concerns a line connecting Chiang Mai, Bangkok, and Phuket in Thailand. In August 2020, TransPod announced the signature of a memorandum of understanding with the government of the province of Alberta, Canada, to support the development of safe, high-speed transportation in Alberta.

TransPod has reportedly secured USD $15M from the Italian industrial investment group Angelo Investment.

Hardt Global Mobility. Hardt Global Mobility was created in Delft, Netherlands, on Sept 9, 2016 by former members of the hyperloop student team from Delft University. The Delft team continues to regularly compete at the SpaceX hyperloop competition, and Hardt is working on the design of a full commercial system. Hardt secured funding and collaborated with industrial partners in the clean energy sector to install a 30m long tube on the campus and run initial tests. They demonstrated a functional system and an important lane-switching technology during a "Grand Reveal" public event in June 2019. The company is now designing and planning the European Hyperloop Center, which will host a 2.6 km test track with a cargo-scale tube of 1.4m diameter, in the Dutch province of Groningen.

Hardt has reportedly raised more than 10M€ (USD $11M) in several rounds, from various investors, including EIT InnoEnergy, the German fund Freigeist Capital, multiple Dutch and Belgium investors, and several parties behind the early growth of Uber.

Zeleros. Zeleros was founded in November 2016 in Valencia, Spain, by three former members of the team Hyperloop UPV, a student team from the University of Valencia. After the success of the student team, winning the "Top Design Concept" and "Best Propulsion Subsystem" awards in 2015, the founders of Hyperloop UPV decided to create Zeleros and pursue the development of a full hyperloop system, apart from the University team. The student team continues to compete regularly at the SpaceX hyperloop pod competition. Zeleros has a stated aim to minimize the cost of infrastructure by integrating levitation and propulsion subsystems into the vehicle rather than the guide way.

Zeleros has secured some partnerships and initial funding, and is planning to begin construction of a 2km long test track in Sagunto, north of Valencia. Zeleros has reportedly received (USD $8.1M) in investment funding.

Nevomo. Nevomo, initially named Hyper Poland, was created in April 2017 in Warsaw, Poland, by members of a student team that took part in the Hyperloop pod competitions from 2015 through 2017. They intend to incrementally develop a hyperloop solution. Their phased, multi-step approach aims to reduce the risk and capital for development of hyperloop lines. First, they propose developing a magnetic propulsion system that could be applied to regular tracks and trains, and then to create a vacuum environment to increase the maximum speed of trains on existing tracks. Next, Nevomo proposes to use subsystems tested in previous phases to build a fully new infrastructure in which a train could reach speeds of up to 1,200km/h.

On October 22, 2019, the company demonstrated its Magrail technology in Warsaw. The test vehicle reached a top speed of about 50 km/h on 48 meters long, 1:5 scale test rail.

Nevomo has reportedly secured PLN 16.5M (USD 4.4M) from the National Center for Research and Development in Poland, and has closed two equity crowdfunding campaigns for an additional 880k€ (USD 375k).

rLoop. rLoop, the only non-student team that has participated in the Hyperloop Pod Competition, was created in June 2015. It defines itself as a crowdsourced engineering company working on several concepts, including flying devices for humans and decentralized prototyping platforms. After having paused their hyperloop developments for a time, in September 2019 rLoop purchased the intellectual property of Arrivo, a company founded by a former Hyperloop One founder that had shut down its operations in 2018.

SwissPod. SwissPod was created in Switzerland in March 2019 by former members of the student team of the Ecole Polytechnique Fédérale de Lausanne (EPFL) participating at the Hyperloop Pod Competitions. The company particularly focuses on developing a line linking Geneva to Zurich, Switzerland. That route was the former focus of the company SwissMetro, created in 2000 to bring a hyperloop-like transportation system to Switzerland. SwissMetro shut down its development activities in 2009, several years prior to the Hyperloop Alpha paper.

Related start-up companies have also formed to support the hyperloop companies. Going beyond the industrial partners who offer to design and manufacture components aside from their regular businesses, these start-ups have the sole purpose of providing key system capabilities. Among them are Continuum Industries which focuses on developing route design software for hyperloop lines, Loop Odyssey which provides operation control systems, and more recently, Hyperloop Italia, created by a founder of HyperloopTT, with plans to conduct feasibility studies in Italy geared toward HyperloopTT's technologies.

Finally, although Elon Musk initially said he does not have time to develop the hyperloop himself, he created The Boring Company in 2016 to reduce the cost of tunneling while improving the speed of digging. Interestingly, one of their commercial applications are tunnels in which automated electric vehicles will transport people at 155mph, 30 feet under cities. The company broke ground in November 2019 for its first contract, a $52.5 million project under the Las Vegas Convention Center. This application, called "Loop," is not – for the moment – hyperloop.

The Public Sector and Hyperloop

The Public sector can play a strong role in making hyperloop a day-to-day reality. Regardless of the innovation within industry and academic departments, hyperloop will not be built and thrive unless governments cooperate and coordinate certain activities. Without clear government regulation and government planning hyperloop would face large barriers to development.

For hyperloop to be viable as a new form of transportation, hyperloop systems must be accepted into existing and new regulatory frameworks, and they must be seen as viable alternatives in master transportation plans. At present, most hyperloop companies are not seeking government funding for the construction or operation of hyperloop lines, but instead envision that public-private partnerships are a viable path forward.

Only six years after the publication of the Alpha paper, many governments, countries, provinces, states, and municipalities have shown strong interest in the opportunities created by ultra-fast, clean ground transportation. This in part has come from the global leadership of the aforementioned companies, who have raised awareness and curiosity about the status and capabilities of hyperloop in the public sector.

Governments have helped in several low-cost but consequential ways. As companies reached the point of needing to build test facilities and tracks, some important partnerships and agreements were established with public authorities. Early interest was expressed from the state of Nevada, USA, which provided incentives and the opportunity for building the Virgin Hyperloop DevLoop test track. In addition, the city of Calgary, Canada issued a resolution to support the creation of a research center to understand the viability of hyperloop technologies. The city of Toulouse and the local region of Haute-Vienne, France, host the test sites of HyperloopTT and TransPod, respectively.

Also, while much of the public sector continued to view tube transportation as a far off, and perhaps impossible, dream, some countries are allocating resources for detailed assessments of the technology. The following table lists some of the more notable actions of governments around the world to understand and prepare for hyperloop.

Netherlands 2017	The Dutch Minister of Infrastructure and the Environment ordered a study to assess the possibilities of building a hyperloop test track in Netherlands.
France July, 2018	The Parliamentary Office for Scientific and Technological Assessment (OPECST) published a briefing assessing current technological developments.
United Kingdom September 2018	Innovate UK, a branch of UK Research and Innovation (national funding agency investing in science and research in the UK) published a report on the opportunities to be created by the hyperloop industry for the UK supply-chain.
Ukraine January, 2019	The National Academy of Sciences produced its own report and delivered it to the Ministry of Infrastructure.
Canada August 2020	The Ministry of Transportation published a study on the feasibility of hyperloop technology.

National (Governmental) Activities to Assess Hyperloop Technology and Viability

It is also noteworthy that Virgin Hyperloop has been selected by the US Department of State to lead the US pavilion at the 2020 World Expo in Dubai (postponed until 2021 because of the Covid-19 pandemic).

The strongest involvement from governments has been the creation of regulatory bodies to prepare for this new mode of transportation. Having regulation in place is a prerequisite for operating any public transport systems and for convincing users that it will be safe and reliable.

In June 2017, DG MOVE, the European Commission's Directorate-General for Mobility and Transport, established a working group with the objective of drafting European standards and a regulatory framework for hyperloop. Four companies (Hardt Global Mobility, TransPod Inc., Zeleros, and Nevomo) became members. This initiative led to the creation of a joint technical committee called the JTC 20. As part of the European Committee for Standardization (CEN) and the European Committee for Electrotechnical Standardization (CENELEC), the goal of this technical committee is to "define, establish, and standardize the methodology and framework to regulate hyperloop travel systems and ensure interoperability and high safety standards throughout Europe."

In March 2019, the U.S. Department of Transportation created the Non-Traditional and Emerging Transportation Technology Council (NETT). This council is "tasked with identifying and resolving jurisdictional and regulatory gaps that may impede the deployment of new technology, such as tunneling, hyperloop, autonomous vehicles, and other innovations." Both actions show that major governments can envision a future with hyperloop and want to be prepared for their role in achieving that vision.

In August 2020, the province of Alberta, Canada, signed a Memorandum of Understanding with TransPod Inc. to study the feasibility of a TransPod hyperloop line in Alberta.

Finally, in December 2020, the European Union explicitly identified hyperloop as a game-changing mobility technology, in the recently published "Sustainable and Smart Mobility Strategy", and "will work towards facilitating testing and trials, and towards making the regulatory environment fit for innovation, so as to support the deployment of solutions on the market."

Hyperloop Routes, Projects, Worldwide Reach

In 2003, Mark Ovenden created an amazing map for the Penguin book "Transit Maps of the World." Modeled after city subway maps, like the schematics of the London Underground, it connects dozens of major cities around the world, even connecting all continents except Antarctica. Today, this map is being used as an example of a potential hyperloop network in a global, connected world. Although such a far-reaching network remains a futuristic vision, the map brings out the need to consider revolutionary changes that could happen in transportation. The world could be connected physically by high-speed ground transportation, much like the way the World Wide Web connects the world electronically.

Each hyperloop company faces several fundamental questions: where would high-speed transportation be most beneficial, how and how much would it directly compete with high-speed rail and airplanes, and what new applications and markets would it create? Indeed, it would not make sense to develop a new means of transportation without identifying its market. These questions closely relate to the investigation of routes. Of course, the first proposed route is the one that inspired the Alpha paper – connecting Los Angeles to San Francisco, California (USA), with expansion to San Diego and Las Vegas, Nevada, where hyperloop was put forward as a better alternative than the California High-Speed Rail project.

Beyond single routes, the hyperloop industry and government planners must consider regions and network connectivity. Neglect in this regard has been noted as contributing to the lack of success of earlier transportation projects, such as the Aerotrain Bertin in France, and the SwissMetro in Switzerland. The economic advantages of networks (discussed in Chapter 5) matter greatly.

Business development for the industry has the task of identifying which pairs of cities or airports would benefit from very high-speed transportation, and which governments would welcome such projects. In a bold move in 2016, Virgin Hyperloop created a global competition for entrants to propose and study possible hyperloop routes. The company issued "a call for comprehensive proposals to build hyperloop networks connecting cities and regions around the world." Amazingly, over 2,600 teams submitted proposals, among which 33 proposals were short-listed, and 10 winning routes were announced in September 2017.

All six major hyperloop companies described earlier have pursued route development, and several have formal agreements with local governments to participate in preliminary feasibility studies. These are taking place in: Alberta (Canada), Abu Dhabi (UAE), Andhra Pradesh, Karnataka, and Maharashtra (India), Slovakia, Germany, Ukraine, the Czech Republic, Indonesia, South Korea, Ohio and Missouri (USA), Saudi Arabia, France, and China.

The novelty and great potential of hyperloop has resulted in an over-abundance of logical routes compared to funding sources available to conduct complete feasibility studies for each of them. For most routes analyzed above, companies have published only initial findings, and for some routes they have kept results confidential for business reasons. Despite companies' public announcements and posts on social media channels, specific estimates of cost, ridership, economic benefit, and other important factors have not been made public for most completed studies. Nevertheless, some detailed reports have recently been released:

- The first fully-public preliminary feasibility study was released early 2019 by TransPod for a line connecting North, Central, and South Thailand.
- In March 2019, the state of Missouri, USA, created a blue-ribbon panel including academics and members of both the public and private sector to study the findings of a Virgin Hyperloop feasibility study on a route connecting Kansas City, Columbia, and St. Louis, Missouri. The panel presented its analysis in October 2019.
- In October 2019, the board of Tampa Bay Area Regional Transit Authority (TBARTA) unanimously decided to allocate $220,000 for a 12 month feasibility study to examine the application of hyperloop, aerial gondolas, and air taxis in the bay. This was funded from a $1M budget allocated in June 2019 by the State legislature of Florida to TBARTA to study and develop transit innovations.
- In early 2018, the Northeast Ohio (USA) Areawide Coordinating Agency (NOACA) entered a public-private partnership to conduct a Great Lakes hyperloop feasibility study. This route would link Chicago, Cleveland, and

Pittsburgh. The report was officially released in December 2019.

- In April 2020, Hardt released a report on a hyperloop connection between Brussels and Amsterdam.

Finally, as of September 2020, perhaps the most realized commercial hyperloop project worldwide consists of a recently-cancelled line between Mumbai and Pune (Maharashtra, India). In July 2019, the government of Maharashtra deemed hyperloop an official "public infrastructure project." The Virgin Hyperloop-DP World Consortium lead the project. The project plan contains two phases, phase 1 being the construction of a demonstration track approximately 11 km long near Pune, and phase 2 being the construction of the remainder of the route to connect these two megacities. The project also included the creation of a regulatory body. Unfortunately, this project has been cancelled by the newly elected government of Maharashtra, stating that the technology, not yet operational anywhere, was considered too risky.

Conclusion

In 2013, Elon Musk rekindled the more than 100-year-old tube transportation industry. Unlike previous transport technologies developed over decades, this newborn hyperloop industry is growing and getting into shape very quickly. From student competitions to commercial projects, it catches lots of attention from both the public and the private sectors, investors, and regulators, with work already underway on the standards and regulations. Climate change issues boost the need for new solutions of ultra-fast and sustainable ground transportation systems. France is currently discussing the possibility of replacing short-haul domestic air routes with high-speed ground connections. This example of a newly-created industry shows how globalization and a connected world can contribute to solving mankind's biggest challenges.

CHAPTER 12. Final Thoughts

Hyperloop creates excitement given its transformative potential as a means to travel quickly and easily over long distances. Whether for daily commutes or transcontinental trips, this mode of transport can use renewable energy sources that integrate more harmoniously with the built environment. The challenges and opportunities facing hyperloop have been presented in this book, while offering a vision of what the future may bring. Its aim has been to contribute to an understanding of hyperloop's true benefits and challenges, one critically needed by societies in order to make good choices about an effort as large and transformational as hyperloop.

Although technological breakthroughs cannot be scheduled, they have appeared regularly throughout our history. They are seldom the result of random luck, but instead spring from the hard work of smart, persistent people open to discovery. They are always built on a great body of existing knowledge. Hyperloop is no different.

As described in previous chapters, engineering challenges must be solved to make hyperloop a reality. Prime examples include the need to balance the size, speed, pressure, and thermal properties of hyperloop to physical constraints. Hyperloop must also address the need to carry sufficient power onboard with batteries, or otherwise to transfer power to a fast-moving pod.

Other key advances already exist, such as developments in magnetic levitation with new materials and with arrays of magnets. Also, one hyperloop company has recently demonstrated a method for a maglev vehicle to switch tracks, a key feature in an operating hyperloop network. Progress has also been made in tunneling technology, whose promise to improve speed and reduce cost has led it to receive more attention than ever before. Although some technological challenges appear daunting, many are being overcome relatively quickly.

Challenges that are less technical than they are design-related have also been discussed above. Safety, of course, is a prime consideration, especially for passenger travel. Although hyperloop will likely be used initially for freight, the safety and operation will be evaluated and improved with use over time. In addition to meeting the ultimate design objective of carrying passengers swiftly and safely, amenities for human travelers also merits attention. This includes mitigating the body's response to accelerations and, as response to the recent pandemic has shown, having a clear focus on health. These issues can all be addressed. Many are already receiving attention from hyperloop companies and from faculty and students at universities around the world.

Another set of challenges sits within the purview of governments and regulators. As described above, these organizations must secure the long, straight right-of-way needed for high speeds. This means that, at least for the first routes developed, hyperloop is likely to be above ground. Governments have a lead role in this, and in forming the essential regulations, rules, and standards that ensure the protection of public interest.

Strong private-academic-public partnerships must be established for the hyperloop concept to be realized and reach its full potential. And of course individuals must decide whether they believe in a future with hyperloop.

This book offers many take-aways, including a better understanding of hyperloop concepts and challenges, and of visions for the future that hyperloop can create. But beyond these, the following:

A world connected by hyperloop networks – the "physical internet" – is not only possible, it is beneficial and within reach.

Resources, commitment, persistence, cooperation, and certainly compromise are essential. Each individual must weigh the benefits against the costs, and against alternatives. This book's authors believe societies must sometimes take a leap of faith and invest in ideas that may fail. Although transformative, the development of such technologies should be done in a staged, thoughtful way. New knowledge should be added as it becomes available to fine-tune the route forward. In the case of hyperloop, and with the insights summarized in this book, we believe the benefits are well worth the effort.

If you are inspired by a future with hyperloop in it, and support the idea that this future should be created based on research, open knowledge, cooperation, and a spirit of public good, then we invite you to become a member of the Hyperloop Advanced Research Partnership (HARP). HARP is an all-volunteer effort. It will benefit from your membership, and even more if you choose to volunteer your time, energy, and passion. Hyperloop can be a great step forward for societies and could become one of the signature accomplishments of our time. If so, with your contribution through HARP you one day in the future may look back and say, "I was a part of that!"

Further Reading

There are many excellent sources of reliable information on topics related to hyperloop. The list below provides a few, as an entry point to delve more into the topics of each chapter. We have strived for a selection that includes both non-technical and moderately-technical content. More technical content can be found in academic literature in areas transportation, various disciplines of physical and engineering science, economics, safety, human factors, and more. The body of research directly applicable to hyperloop is growing quickly. Also, entries for Chapter 11 include many links to companies and groups actively developing hyperloop products and services.

On hyperloop background and general topics

Musk, Elon, 2013. Hyperloop Alpha, SpaceX, available at https://www.spacex.com/sites/spacex/files/hyperloop_alpha-20130812.pdf (accessed May 24, 2019).

Swartzwelter, B. (2003). *Faster than Jets*. Kingston, WA: Alder Press.

Taylor, C. L., Hyde, D. J., & Barr, L. C. (2016). *Hyperloop commercial feasibility analysis: high level overview* (No. DOT-VNTSC-NASA-16-01). John A. Volpe National Transportation Systems Center (US).

Decker, K, et. al., 2017, Conceptual Feasibility Study of the Hyperloop Vehicle for Next-Generation Transport, presented at the American Institute of Aeronautics and Astronautics, Reston, VA. https://ntrs.nasa.gov/archive/nasa/casi.ntrs.nasa.gov/20170001624.pdf

Janzen, R., 2017: TransPod ultra-high-speed tube transportation: dynamics of vehicles and infrastructure. Procedia Engineering, 199, 8-17. https://doi.org/10.1016/j.proeng.2017.09.142 (open access)

van Goeverden, K., D. Milakis, M. Janic, and R. Konings, 2018: Analysis and modelling of performances of the HL

(Hyperloop) transport system. European Transport Research Review, 10(2), 41. https://doi.org/10.1186/s12544-018-0312-x (open access)

Abdelrahman, A. S., Sayeed, J., & Youssef, M. Z. (2017). "Hyperloop transportation system: analysis, design, control, and implementation." *IEEE Transactions on Industrial Electronics*, 65(9), 7427-7436.

Chapter 1

Egli, Dane (2020). Public-Private Partnerships & Natural Hazards Governance. Oxford Encyclopedia of Natural Hazards Governance. January 2020, Oxford University Press: London.

Molendijk, Koen (2017). The Development of Value Propositions in High-Tech Entrepreneurial Startups. University of Twente, The Netherlands. Web of Science.

Murphy, Gregory; Tocher, Neil; Ward, Bryant (2016). An Examination of Public, Private, Academic Partnerships. Public Organization Review: December 2016.

Ostrom, Elinor (1990). Governing the Commons: The Evolution of Institutions for Collective Action. Cambridge University Press: Cambridge. ISBN 978-0-52140599-7.

Patala, Samuli (2016). Sustainable Value Propositions: Framework and Implications for Technology Suppliers. Industrial Marketing Management, Volume 59, November 2016, pp 144-156.

Chapter 2

Marchetti, C. (1994). Anthropological Invariants in Travel Behavior, Technological Forecasting and Social Change, 47, 75-88.
http://www.cesaremarchetti.org/archive/electronic/basic_instincts.pdf

Donaldson, D., and R. Hornbeck (2013), Railroads and American Economic Growth: A "Market Access" Approach, NBER Working Paper Series, Working Paper 19213, National Bureau of Economic Research, July 2013.
https://www.nber.org/papers/w19213

David, Paul A. 1969. "Transport Innovation and Economic Growth: Professor Fogel on and off the Rails." Economic History Review, 22(3): 506–524.
https://onlinelibrary.wiley.com/doi/abs/10.1111/j.1468-0289.1969.tb00186.x

Capka, J. R., (2006), Celebrating 50 Years: The Eisenhower Interstate Highway System, Hearing on Celebrating 50 Years: The Eisenhower Interstate Highway System, June 27, 2006, https://www.transportation.gov/testimony/celebrating-50-years-eisenhower-interstate-highway-system

Weingroff, Richard F., 1996. "Federal-Aid Highway Act of 1956, Creating the Interstate System". Public Roads. 60 (1). ISSN 0033-3735.

Negroni, C., (2019), The Flight that Changed Everything,, When the 707 gave us the world, Air & Space Magazine, February 2019, https://www.airspacemag.com/history-of-flight/707-flight-changed-everything-180971219/

Wilson, Mark, It's time to redesign travel for the age of COVID-19, Fast Company 03-11-20. https://www.fastcompany.com/90474397/its-time-to-redesign-travel-for-the-age-of-covid-19 (accessed April 19, 2020).

Juergen T Steinmetz, Juergen T., How Airlines could stop the spread of COVID19 on a plane? Invest $1!, eTurboNews, February 21, 2020, https://www.eturbonews.com/543573/how-airlines-could-stop-the-spread-of-covid19-on-a-plane/ (accessed April 19, 2020).

Chapter 3

Max M. J. Opgenoord, Philip C. Caplan: On the Aerodynamic Design of the Hyperloop Concept, 35th AIAA Applied Aerodynamics Conference, http://web.mit.edu/mopg/www/papers/OpgenoordCaplan_2017_Aerodynamics_Hyperloop_online.pdf

Prasad, N., Jain, S. & Gupta, S. Electrical Components of Maglev Systems: Emerging Trends. Urban Rail Transit 5, 67–79 (2019). https://doi.org/10.1007/s40864-019-0104-1 https://link.springer.com/article/10.1007/s40864-019-0104-1

The Importance of Aerodynamics in a Near Vacuum, Delft Hyperloop https://hyperloopconnected.org/2019/03/the-importance-of-aerodynamics-in-a-near-vacuum/

U.S. Department of Energy, How Maglev Works, June 14, 2016, https://www.energy.gov/articles/how-maglev-works

Wilson, Cornell, Maglev: Magnetic Levitating Trains, https://sites.tufts.edu/eeseniordesignhandbook/2015/maglev-magnetic-levitating-trains/

Santora, Mike, 2016, Comparing the Real Costs of Vacuum Generators, https://www.pneumatictips.com/comparing-the-real-costs-of-vacuum-generators/

Maglev.net, 2018, The Six Operation Maglev Lines in 2018, https://www.maglev.net/six-operational-maglev-lines-in-2018

Chapter 4

TIME Magazine, See the Futuristic Pods That Could Change How We Travel. Time Photo February 4, 2016.

Hale et al. 2011. Wings in Orbit; Scientific and Engineering Legacies of the Space Shuttle 1971-2010. Editors: Wayne

Hale, Helen Lane, Gail Chapline, Kamiesh Lulla. National Aeronautics and Space Administration. 2011. NASA

https://www.nasa.gov/centers/johnson/pdf/584739main_Wings-ch5d-pgs370-407.pdf

Harris LR, Jenkin M, Dyde RT, Jenkin H. 2011. Enhancing visual cues to orientation: suggestions for space travelers and the elderly. Eds: AM Green, CE Chapman JF Kalaska and F Lepore. In: Progress in Brain Research, Elsevier B.V. ISSN: 0079-6123.

Harris LR, Herpers R, Hofhammer T, Jenkin M. 2014. How much gravity is needed to establish the perceptual upright? PLoS ONE 9(9): e106207. Doi: 10.1371/journal.pone.0106207.

Yakushin SB, Martinelli GP, Raphan T, Cohen B. 2016. The response of the vestibulosympathetic reflex to linear acceleration in the rat. J. Neurophysiol. 116(6): 2752-2764.

Chapter 5

Kim, Iljoong, Hojun Lee, and Ilya Somin, eds., 2017: Eminent Domain: A Comparative Perspective, Cambridge University Press, 328 pp., ISBN-10: 1316628337, ISBN-13: 978-1316628331.

Heiner, Jared D. and Kara M. Kockelman, 2005: Costs of Right-of-Way Acquisition: Methods and Models for Estimation, J. Transpor. Eng., 131(3), https://doi.org/10.1061/(ASCE)0733-947X(2005)131:3(193).

Diaz, R. B., 2005: Impacts of Rail Transit on Property Values, Proc. 1999 Commuter Rail/Rapid Transit Conference, Toronto, CA, http://www.rtd-fastracks.com/media/uploads/nm/impacts_of_rail_transif_on_property_values.pdf.

Maidl, B., M. Thewes, and U. Maidl, 2013, Handbook of Tunnel Engineering, Volume I: Structures and Methods, Wilhelm Ernst & Sohn, Wiley, Germany.

Nickelsburg, J., S. Ahluwalia, and Y. Yang, 2018: High-Speed Rail Economics, Urbanization and Housing Affordability Revisited: Evidence from the Shinkansen System, https://www.anderson.ucla.edu/documents/sites/faculty/re view%20publications/research/Nickelsburg_High-Speed_Rail_Economics__Urbanization_and_Housing_Affor dability_2018.pdf

Chapter 6

Cramtion, P., R. R. Geddes, and Axel Ockenfels. 2019. "Using Technology to Eliminate Traffic Congestion," Journal of Institutional and Theoretical Economics, 175:1, 126-139.

Geddes, R. R., The Road to Renewal: Private Investment in U.S. Transportation Infrastructure (Washington, DC: AEI Press) 2011.

National Academies of Sciences, Engineering, and Medicine. 2019. Leveraging Private Capital for Infrastructure Renewal. Washington, DC: The National Academies Press. https://doi.org/10.17226/25561.

Taylor, C. L., Hyde, D. J., & Barr, L. C. (2016). Hyperloop commercial feasibility analysis: high level overview (No. DOT-VNTSC-NASA-16-01). John A. Volpe National Transportation Systems Center (US).

Abdelrahman, A. S., Sayeed, J., & Youssef, M. Z. (2017). "Hyperloop transportation system: analysis, design, control, and implementation." IEEE Transactions on Industrial Electronics, 65(9), 7427-7436.

Chapter 7

Sutton, Ian (Ed), Process Risk and Reliability Management (Second Edition), Gulf Professional Publishing, 2015. ISBN 9780128016534

Chapter 8

International Civil Aviation Organization. ICAP Environmental Report 2010: Aviation and Climate Change. Montreal: ICAO, 2010.

International Energy Agency. The Future of Trucks, Implications for energy and the environment, Second Edition. Paris: IEA, 2017.

International Energy Agency. The Future of Rail, Opportunities for Energy and the Environment. Paris: IEA, 2019.

Penner, J. E., D. H. Lister, D. J. Griggs, D. J. Dokken, M McFarland, and (Eds). "IPCC Special Report: Aviation and the Global Atmosphere." Intergovernmental Panel on ClimateChange. 1999. https://www.ipcc.ch/report/aviation-and-the-global-atmosphere-2/ (accessed May 19, 2019).

Schlossberg, Tatiana. Flying Is Bad for the Planet. You Can Help Make It Better. July 27, 2017. https://www.nytimes.com/2017/07/27/climate/airplane-pollution-global-warming.html] (accessed May 19, 2019).

Uddin, Waheed, Patrick Sherry, and Burak Eksioglu. Integrated Intermodal Transportation Corridors for Economically Viable and Safe Global Supply Chain. National Center for Intermodal Transportation for Economic Competitiveness, 2016.

Unger, N. (2009). Transportation Pollution and Global Warming. Retrieved from NASA Goddard Institute for Space Studies: https://www.giss.nasa.gov/research/briefs/unger_02/

Chapter 9

M. Bierlaire, A. de Palma, R. Hurtubia, P. Waddell (Eds.) 2015. Integrated Transport & Land Use Modeling for Sustainable Cities, EPFL Press.

https://hyperloopconnected.org/2018/04/societal-impact-of-the-hyperloop/ (retrieved December 14, 2019)

https://www.inverse.com/article/35274-hyperloop-one-urban-development-reddit (retrieved December 14, 2019)

Denver Department of Environmental Health. (2014). How neighborhood planning affects health in Globeville and Elyria-Swansea. Denver, CO. Accessed November 30, 2019. https://www.denvergov.org/content/dam/denvergov/Portals/746/documents/HIA/HIA%20Composite%20Report_9-18-14.pdf

Ratner, Keith A. and Andrew R. Goetz, 2013: The reshaping of land use and urban form in Denver through transit-oriented development, Cities, Vol. 30, p. 31-46, https://doi.org/10.1016/j.cities.2012.08.007

Porter, Michael E., 1998: Clusters and the New Economics of Competition, Harvard Business Review, Nov-Dec 1998.

Chapter 10

Egli, Dane S., 2014. "Beyond the Storms: Strengthening Homeland Security and Disaster Management to Achieve Resilience." Routledge/Taylor & Francis: New York

Limani, Ylber, 2016. "Applied Relationship between Transport and Economy." Science Direct Conference Paper. https://www.sciencedirect.com/science/article/pii/S2405896316324960X.

Sherwood, Christina H, 2010. "For vibrant economy, transportation networks matter more than city size." ZDNet Innovation, September 21, 2010, London, United Kingdom.

https://www.zdnet.com/article/for-vibrant-economy-transportation-networks-matter-more-than-city-size/

Benny, Daniel J., 2015. "Maritime Security: Protection of Marinas, Ports, Small Watercraft, Yachts, Ships." Boca Raton, Florida: Taylor & Francis Group

Department of Transportation (DOT), 2012. "Freight Flows by Highway, Railroad, and Waterway: 2012." Bureau of Transportation Statistics, Freight Flow Map. https://www.bts.gov/content/freight-flows-highway-railroad-and-waterway-2012

Department of Transportation (DOT), 2017. "Freight Shipments by Domestic Mode, Bureau of Transportation Statistics and Federal Highway Administration, Freight Analysis Framework.

Department of Transportation (DOT), 2017. "A Journey Through American Transportation: 1776 – 2017." DOT 50th Anniversary Article, Washington, DC. https://www.transportation.gov/50/timeline

Global Facilitation Partnership (GFP), 2018. Global Facilitation Partnership for Transportation and Trade. "Maritime Transport and Port Operations."

https://www.usni.org/magazines/proceedings/2019/january/breaking-faith-americas-coast-guard.

United Nations Conference on Trade and Development (UNCTAD), 2018. "Review of Maritime Transport Annual Report." New York and Geneva. https://unctad.org/en/PublicationsLibrary/rmt2018_en.pdf

Zukunft, Paul, 2019. "Breaking Faith with America's Coast Guard." U.S. Naval Institute Proceedings, Annapolis, MD. January 2019, Vol. 145/1/1,391.

Friedman, Thomas L., 2014. "Makers and Breakers." New York Times Sunday Review, Opinion Editorial, November 8, 2014. https://www.nytimes.com/2014/11/09/opinion/sunday/thomas-l-friedman-makers-and-breakers.html

Global Resilience Institute (GRI), 2019. Definition of Resilience from Presidential Policy Directive (PPD-21). https://globalresilience.northeastern.edu/about/overview/

Organization for Economic Cooperation & Development (OECD), 2018. "Transformative Technologies and Jobs of the Future." Background report for the Canadian G7 Innovation Ministers' Meeting Montreal, Canada, March 27-28, 2018.

Browne, Malcolm W., 1991. "Invention That Shaped the Gulf War: Laser-Guided Bomb." The New York Times. February 26, 1991. https://www.nytimes.com/1991/02/26/science/invention-that-shaped-the-gulf-war-the-laser-guided-bomb.html

Coletta, Damon V., 2018. "Navigating the Third Offset Strategy," Learning from Military Transformations.

Garthoff, Raymond L., 2007. "Why Did the Cold War Arise, and Why Did It End?" Oxford Academic Diplomatic History, Volume 16, Issue 2, pages 287–293. https://academic.oup.com/dh/article-abstract/16/2/287/338743

Chapter 11

HyperloopTT, https://www.hyperlooptt.com/

Virgin Hyperloop, https://virginhyperloop.com/

TransPod Inc., https://transpod.com

Hardt Global Mobility, https://hardt.global

Zeleros, https://zeleros.com

Nevomo, https://www.nevomo.tech/en/

rLoop, https://www.rloop.org

SwissPod, https://swisspod.ch

Continuum Industries, https://continuum.industries

Loop Odyssey, http://loop-odyssey.com

The Boring Company, https://www.boringcompany.com

Hyperloop Italia, https://hyperloopitalia.com/

U.S. Department of Transportation, Non-Traditional and Emerging Transportation Technology Council (NETT), https://www.transportation.gov/nettcouncil

Overden, Mark, Transit Maps of the World, Penguin, 2015, 176 pp. ISBN 9780141981444, https://www.penguin.co.uk/books/286645/transit-maps-of-the-world/9780141981444.html

Hyperloop One Global Challenge, 2017, https://hyperloop-one.com/global-challenge

A newcomer in the European transport standardization family: JTC 20 on hyperloop systems. 2020-02-05. https://www.cencenelec.eu/news/articles/Pages/AR-2020-003.aspx

TransPod, March 2019, Hyperloop in Thailand, Preliminary study on the implementation of a TransPod Hyperloop line in Thailand, 118 pp., https://transpod.com/wp-content/uploads/2019/03/Final_Report_TransPod_Hyperloop_Thailand.pdf

Tampa Bay Regional Transit Authority (TBARTA), Record Level $2.5 Million in TBARTA Funding Approved to Advance Regional Transit in Tampa Bay, https://www.tbarta.com/en/about/news-media-releases/2019/record-level-25-million-in-tbarta-funding-approved-to-advance-regional-transit-in-tampa-bay/

Northeast Ohio (USA) Areawide Coordinating Agency (NOACA), NOACA Board awards contract for Great Lakes Hyperloop Feasibility Study, June 12, 2018, https://www.noaca.org/Home/Components/News/News/11824

Pune Metropolitan Regional Development Authority, Public Declaration, 2018, http://pmrda.gov.in/Marathi/images/PDF's/Public%20Declaration%20English.pdf

The Indian Hyperloop Pod Competition
https://ihpc2020.web.app/

Figure Credits and Information

p. 5   High speed tube traveling technology concept.

        Credit: Shutterstock

p. 8   Toamasina commercial port (Madagascar)

        Credit: Thierry Boitier

p. 8   Singapore skyline

        Credit: Shutterstock

p. 9   The original "Champagne Photo" taken May 10, 1869 at the joining of the Union Pacific track.

        Credit: National Park Service website. https://www.nps.gov/gosp/learn/historycultur e/a-moment-in-time.htm

p. 9   Major Deegan expressway seven miles outside New York City has six lanes to accommodate increasing suburban commuter traffic in 1957.

        Credit: Shutterstock

p. 9   1958, Seattle, Washington. The first three 707-121 aircraft with Pratt&Whitney JT3C6 engines awaiting delivery to Pan American World Airways.

        Credit: Wikimedia Commons

p. 13   August 18, 2018, Varreddes, France. A SNCF TGV Duplex high speed train.

        Credit: Shutterstock

p. 30   Hyperloop tube illustration

        Credit: TransPod

p. 32   Pod interior concept.

        Credit: Joakim Sandén and TransPod

p. 33   Pod interior concept.

        Credit: Hardt-Plompmazes

Credit: John McColgan, Wikimedia commons

p. 74   Drought-19478-1920.jpg.

Credit: Image by PublicDomainPictures from Pixabay

p. 74   Wind farm at dawn.

Credit: Pixabay. https://www.pexels.com/photo/afterglow-alternative-energy-clouds-dawn-532192/

p. 76   Wheel-to-well carbon intensities. Data from The Future of Rail, International Energy Agency 2019.

Credit: The authors

p. 76   Wind farm at dawn.

Credit: Pixabay.

p. 77   Global CO_2 emission by sector and by transport mode. Data from International Energy Agency.

Credit: The authors

p. 78   Diesel truck emitting soot.

Source: Environmental Protection Agency, Wikimedia commons https://commons.wikimedia.org/wiki/File:Diesel-smoke.jpg

p. 79   Passenger cars in smog

Source: Environmental Protection Agency website. https://www.epa.gov/transportation-air-pollution-and-climate-change

p. 80   Airplane exhaust at takeoff.

Source: Image by stuartcollins1, from Pixabay. https://pixabay.com/photos/takeoff-plane-aircraft-travel-2678096/

p. 81   Estimated distribution of number of US aircraft flights by leg distance (stage length), and number of miles flown by stage length. Data from Marien et al. 2018.

Source: The authors.

p. 82   Diesel locomotive emitting soot.

Source: Image by WikiImages from Pixabay. https://pixabay.com/photos/locomotive-diesel-russia-train-60539/

p. 83   Energy intensity by mode of transport for passengers and freight. Data source The future of Rail, International Energy Agency, 2019.

Source: The authors

p. 84   Container ship emitting exhaust.

Credit: Image by Michael Zöllner, from Pixabay

p. 85 CO_2 emissions potential reduction with hyperloop.

Credit: The authors

p. 89   Hyperloop station and tubes, exterior concept.

Credit: REC Architecture, TransPod

p. 90   Wildlife corridor (Ecoduct) crossing at Dwingelderveld National Park, Beilen, Drenthe, The Netherlands.

Credit: Shutterstock

p. 93   Representation of the life-cycle of compounds. Credit: Environmental Protection Agency "Risk Management Sustainability Technology"

p. 95   Hyperloop station interior concept. Credit: Student work from the 2014-15 IDEAS Mobility Studio: Hyperloop with Craig Hodgetts, Marta Nowak and David Ross. Image courtesy of UCLA Architecture and Urban Design

About the Authors

Dario Bueno-Baques

Dr. Dario Bueno-Baques is research physicist in the Department of Physics and Energy Science at the University of Colorado, Colorado Springs, and is on the Board of Directors of the Hyperloop Advanced Research Partnership. His areas of expertise include magnetic and ferroelectric materials and nanostructures, microwave and millimeter-wave devices, electromagnetic systems, and multiphysics simulation. He has authored more than 50 peer-reviewed papers, patents, book chapters, and essays.

Thierry Boitier

With two Masters degrees in both engineering and business management, Thierry Boitier helps shape the strategy, supply chain, and business development of TransPod. He is one of the first employees of that company, and has presented hyperloop and TransPod technologies during conferences and seminars in Canada, the USA, France and the UK. Mr. Boitier has more than a decade of experience in international business and supply chain management, in Europe, Africa, Asia, and North-America. He is also an accomplished musician and a marathon runner.

Radu Cascaval

Professor Radu C. Cascaval is an applied mathematician at the University of Colorado, Colorado Springs. His research includes areas as diverse as human physiology, fluid dynamics, and efficient transportation. Dr. Cascaval has been advising teams of students participating in the SpaceX Hyperloop Pod Competition since 2015. He has been a Senior Advisor to Hyperloop Advanced Research Partnership since 2017.

Stephen A. Cohn

Dr. Stephen A. Cohn leads research with a focus on applying the best technologies to address meaningful societal challenges. For more than two decades Dr. Cohn was a scientist and group leader at the National Center for Atmospheric Research. He is a co-founder and past president of Hyperloop Advanced Research Partnership (HARP), serves on World Meteorological Organization committees, and is the author of more than 45 peer-reviewed scientific papers.

Dane S. Egli

Dr. Dane Egli served in the White House, National Security Council staff and was senior advisor to the President. He also led programs for the National Nuclear Security Administration at Los Alamos National Laboratory. Dr. Egli is a co-founder and former president of Hyperloop Advanced Research Partnership (HARP). He is author of the book, "Beyond the Storms: Strengthening Homeland Security and Disaster Management to Achieve Resilience."

R. Richard Geddes

Professor R. Richard Geddes is a Professor at Cornell University and a founding director of the Cornell Program in Infrastructure Policy (CPIP), as well as a founding director of the Hyperloop Advanced Research Partnership (HARP). He is author of the book, "The Road to Renewal: Private Investment in the U.S. Transportation Infrastructure."

Mary Ann Ottinger

Professor Mary Ann Ottinger has a long history of multidisciplinary research, teaching, mentoring and administration. She is internationally recognized for her research in neuroendocrine systems, development and aging, and has mentored more than 50 graduate students and postdocs. She serves on working groups addressing issues in transportation, conservation, and the interdependence of human and natural systems. Dr. Ottinger served as Associate Vice President/Associate Vice Chancellor for Research at the University of Houston. She has been a Senior Advisor to the Hyperloop Advanced Research Partnership (HARP) since 2016.

Ian Sutton

Ian Sutton is a chemical engineer with over 30 years of design and operating experience in process safety and risk management. He is the owner of Sutton Technical Books and has authored numerous books and ebooks on Safety Management. Mr. Sutton is a senior advisor to the Hyperloop Advanced Research Partnership.

Brad Swartzwelter

Brad Swartzwelter is the author of "Faster than Jets," a book dedicated to examining high speed, low pressure tube transportation years before the word hyperloop was coined. He is President of the Hyperloop Advanced Research Partnership (HARP) and by-day is an Amtrak Conductor.

A Brief Overview of Time and Events for HARP and Acknowledgement of Multidisciplinary Colleagues

When **Dave Clute & John Whitcomb** visited the Hyperloop One test site outside Las Vegas, Nevada in March 2016, they never dreamed that a year later they would be in Washington, D.C. on the short-list of finalists for the **Hyperloop Global Challenge**. This journey began with the founding of the **Rocky Mountain Hyperloop Consortium** (RMHLC) and subsequently, the founding of the **Hyperloop Advanced Research Partnership** (HARP) in March 2017.

Early contributors and members of HARP include **Tracy Hughes** (Silicon Valley Sports Ventures), **Chris Zahas** (Leland Consulting Group), **Brian Donohue** (Johns Hopkins University Applied Physics Laboratory (APL), and **David W. Hewett** (Olive Real Estate Group). There are many others, too numerous to mention in this space, but their contributions are known and appreciated.

The inaugural symposium of HARP was held in Denver, Colorado in March 2017. This coincided with the original team coming together to complete the submission for the Hyperloop Global Challenge. As more members joined the group, HARP grew quickly and our first **Board of Directors** began the work ahead.

The embryonic hyperloop industry will recognize, in the years to come, the important and historic contributions made by the authors of this book, and more importantly visionary transportation practitioners which rely on multidisciplinary scientists, engineers, and trans-disciplinary fields bringing together social sciences, education, and communities.

Made in the USA
Monee, IL
09 March 2023